CO_2 Sequestration

Edited by Leidivan Almeida Frazão,
Adriana Marcela Silva-Olaya
and Junio Cota Silva

Published in London, United Kingdom

IntechOpen

Supporting open minds since 2005

CO_2 Sequestration
http://dx.doi.org/10.5772/intechopen.77533
Edited by Leidivan Almeida Frazão, Adriana Marcela Silva-Olaya and Junio Cota Silva

Contributors
Hesham G. Ibrahim, Mohamed A. Al-Meshragi, Kristofor R. Brye, Marya McKee, Lisa Wood, Askia M. Mohammed, James S. Robinson, David J. Midmore, Anne Verhoef, Nediljka Gaurina-Međimurec, Karolina Novak Mavar, Mohammad Ibrahim Khalil, Rosa Francaviglia, Beverley Henry, Katja Klumpp, Peter Koncz, Mireia Llorente, Beata Emoke Madari, Miriam Muñoz-Rojas, Rainer Nerger, Sikander Khan Tanveer, Xingli Lu, Shamim-Ul-Sibtain Shah, Imtiaz Hussain, Muhammad Sohail, Leidivan Almeida Frazão, Guilherme S. Raucci, João Luis Carvalho, Marcelo Valadares Galdos, Cindy S. Moreira, Carlos Eduardo P. Cerri, Carlos Cerri, Junio C. Silva, Adriana M. Silva-Olaya

First published in London, United Kingdom, 2020 by IntechOpen
IntechOpen is the global imprint of INTECHOPEN LIMITED, registered in England and Wales, registration number: 11086078, 7th floor, 10 Lower Thames Street, London, EC3R 6AF, United Kingdom
Printed in Croatia

British Library Cataloguing-in-Publication Data
A catalogue record for this book is available from the British Library

Additional hard and PDF copies can be obtained from orders@intechopen.com

CO_2 Sequestration
Edited by Leidivan Almeida Frazão, Adriana Marcela Silva-Olaya and Junio Cota Silva
p. cm.
Print ISBN 978-1-83962-992-1
Online ISBN 978-1-83962-993-8
eBook (PDF) ISBN 978-1-83962-994-5

For EU product safety concerns:
IN TECH d.o.o., Prolaz Marije Krucifikse Kozulić 3, 51000 Rijeka, Croatia,
info@intechopen.com or visit our website at intechopen.com.

Meet the editors

Leidivan Almeida Frazão is Professor of Soil Science and coordinator of the Graduate Program in Plant Science (2017-2020) at Federal University of Minas Gerais, Brazil, Institute of Agricultural Sciences. She received a BSc in Agronomic Engineering from the Federal University of São Carlos in 2004, and an MSc in Chemistry in Agriculture and Environment and a DSc in Soil and Plant Nutrition from the University of São Paulo, Brazil, in 2007 and 2012, respectively. She has published papers in international peer-reviewed journals about the management of soil organic matter compartments, soil carbon stocks and greenhouse gas emissions from different agricultural systems. Currently, her main focus of research is dynamics of soil organic matter under agrosilvopastoral and agroforestry systems in Brazil in collaboration with various national and international institutions.

Junio Cota Silva is Professor of Biotechnology and Microbiology at Federal University of Minas Gerais, Brazil, Institute of Agricultural Sciences. He obtained a BSc in Biological Sciences from the Federal University of Viçosa in 2007, and an MSc and DSc in Food Science from the State University of Campinas, Brazil, in 2009 and 2013, respectively. He has more than 15 years' experience in scientific research and has led several research projects on industrial and applied biotechnology and has been awarded research project grants. He has published more than thirty papers on industrial biotechnology and biorenewables in international peer-reviewed journals in collaboration with various national and international institutions worldwide. He is an editorial board member and ad hoc reviewer for international journals including *Biocatalysis*, *Agricultural Biotechnology*, and *BioMed Research International*.

Adriana Marcela Silva-Olaya is Assistant Professor in the Faculty of Agricultural Sciences and coordinator of the Biogeochemical Process Laboratory at the University of the Amazon, Colombia. A major focus of her work involves impact of agricultural management on soil quality and soil C dynamics in the Amazon region. Previous and current research activities include field measurement and modeling of soil C sequestration and greenhouse gas emissions from soils. Currently, Dr. Silva-Olaya is leading a project involving changes in soil microbiome composition and functionality due to transition from forest to pasture in the Colombian Amazon region.

Contents

Preface

In recent decades there has been growing concern over the impact of greenhouse gas (GHG) emissions on global warming and climate changes. Fossil fuel use, deforestation, intensive livestock farming, use of synthetic fertilizers and industrial processes have been pointed out as the main human sources of GHG emissions. As population growth around the world is contributing to this global warming, several efforts have been made to mitigate GHGs by agricultural practices and industrial processes.

The Intergovernmental Panel on Climate Change (IPCC) has reported different strategies to promote atmospheric CO_2 sequestration. However, as the methodologies and new strategies are constantly updated, the dissemination of this information becomes important within the current scenario of climate change.

As such, this book provides the reader with a comprehensive overview of the current state of the art of the strategies that contribute to reducing GHG emissions and promoting CO_2 sequestration. Chapters broadly discuss alternatives such as soil carbon sequestration through conservationist management systems in agriculture, improvement of industrial processes and reuse of residues, adsorption processes that can be performed using activated carbon and effective methods of carbon capture and storage such as geological sequestration of CO_2. The book is divided into eight chapters written by thirty-five researchers in different fields and organized by subjects related to CO_2 sequestration in several regions of the world. The results presented here are directly from the authors' research as well as from studies in important publications.

This book contains relevant information about atmospheric CO_2 sequestration and contributes to discussions on reducing the impacts of global warming and climate change. We recommend this reference to the general public, undergraduate and graduate students, and researchers who aim to deepen their knowledge on the topics discussed within.

Leidivan Almeida Frazão and Junio Cota Silva
Universidade Federal Minas Gerais,
Brazil

Adriana Marcela Silva-Olaya
Universidad de la Amazonia,
Colombia

Chapter 1

Introductory Chapter: CO_2 Sequestration

Leidivan A. Frazão, Junio C. Silva and Adriana M. Silva-Olaya

1. Introduction

The Special Report from the Intergovernmental Panel on Climate Change (IPCC) [1] revealed that recent trends in greenhouse gas (GHG) emissions and the level of international ambition indicated by nationally determined contributions, within the Paris Agreement, deviate from a track consistent with limiting warming to well below 2°C. This will require a drastic reduction in greenhouse gas emissions by 2030 and thereafter removal of carbon from the atmosphere in large quantities. The IPPC reports found that many climate models can only meet the two-degree Celsius goal when carbon removal strategies are included among the potential policy options.

There are several strategies to promote carbon dioxide (CO_2) sequestration by agriculture and industry. So, it is necessary to evaluate the methodologies that have been used and to understand the gaps to achieve more sustainable production systems.

In agriculture, the management of agricultural systems that promote soil carbon sink depends on depth, clay content and mineralogy, plant available water holding capacity, nutrient reserves, landscape position, and the antecedent SOC stock [2]. As the soil carbon fluxes vary according to environmental and anthropogenic driving factors [3], soil carbon sequestration can be a short-term solution of reducing CO_2 concentration in the atmosphere.

In addition to agronomic practices, several effective methods of carbon capture and storage (CCS) have been proposed to reduce the amount of emitted CO_2 in the atmosphere. Adsorption processes can be performed using activated carbon [4] where the adsorptive process can use adsorbents derived from low-cost agro-wastes. Another way to reduce CO_2 emission into the atmosphere is by capturing CO_2 from the flue gases and storing that in deep geological formations [5]. The CCS provides financial offsets in terms of CO_2 sequestration cost.

Therefore, this book provides a comprehensive overview of the current state of the art about the strategies that contribute to reducing GHG emissions and promote CO_2 sequestration by agricultural techniques and carbon capture and storage.

2. Opportunities and challenges for CO_2 sequestration

Several studies have indicated the storage in biomass, soils, adsorption processes, and geological formations as viable techniques for CO_2 sequestration. All these technologies have the potential to mitigate global warming and climate change [2–6].

IntechOpen

Improving agricultural land management techniques is an efficient way to increase carbon uptake and storage. Strategies to ensure soil carbon sequestration can be obtained through the adoption of different agronomic management practices [2]. Land use with grassland species can also maintain and increase soil organic carbon storage over time [7]. Other studies have reported that land use with perennial crops can also be adopted to promote CO_2 sequestration in biomass, and soil is the main component storing the highest amount of carbon in these agroecosystems [8, 9].

Carbon removal can also be achieved through the technology of adsorption on activated carbon from low-cost raw material. Agricultural and forestry residues or biomass residue wastes could be used as suitable raw materials for the production of activated carbon [10]. Furthermore, CCS by geological sequestration is another technological form for carbon removal and can be applied to different industries [5].

3. Perspectives

As the population is growing around the world and indirectly contributes to global warming, several efforts have been made to mitigate GHG emissions. So, the adoption of CO_2 sequestration technologies in the agricultural and industrial sectors has become essential to reduce the impacts of global warming and climate change.

Author details

Leidivan A. Frazão[1]*, Junio C. Silva[1] and Adriana M. Silva-Olaya[2]

1 Universidade Federal de Minas Gerais, Montes Claros, MG, Brazil

2 Universidad de la Amazonia, Florencia, Colombia

*Address all correspondence to: lafrazao@ufmg.br

References

[1] International Panel on Climate Change (IPCC). In: Masson-Delmotte V, Zhai P, Pörtner HO, Roberts D, Skea J, Shukla PR, et al., editors. Global Warming of 1.5°C. An IPCC Special Report on the Impacts of Global Warming of 1.5°C above Pre-Industrial Levels and Related Global Greenhouse Gas Emission Pathways, in the Context of Strengthening the Global Response to the Threat of Climate Change, Sustainable Development, and Efforts to Eradicate Poverty. 2018. Available from: https://www.ipcc.ch/site/assets/uploads/sites/2/2019/06/SR15_Full_Report_High_Res.pdf

[2] Lal R. Digging deeper: A holistic perspective of factors affecting soil organic carbon sequestration in agroecosystems. Global Change Biology. 2018;**24**:3285-3301

[3] Stockmann U, Adams M, Crawford JW, Field DJ, Henakaarchchia N, Jenkins M, et al. The knowns, known unknowns and unknowns of sequestration of soil organic carbon. Agriculture, Ecosystems and Environment. 2013;**164**:80-90

[4] Mohammad SS, Wan Mohd Ashri WD, Amirhossein H, Ahmad S. A review on surface modification of activated carbon for carbon dioxide adsorption. Journal of Analytical and Applied Pyrolysis. 2010;**89**:143-151

[5] Mosleh MH, Sedighi M, Babaei M, Turner M. Geological sequestration of carbon dioxide. In: Letcher TM, editor. Managing Global Warming: An Interface of Technology and Human Issues. London, United Kingdon: Academic Press (Elsevier); 2019. pp. 487-500

[6] Vaughan NE, Gough N, Mander S, Littleton EW, Welfle A, Gernaat DEHJ, et al. Evaluating the use of biomass energy with carbon capture and storage in low emission scenarios. Environmental Research Letters. 2018;**13**:044014. DOI: 10.1088/1748-9326/aaaa02

[7] Hungate BA, Barbier EB, Ando AW, Marks SP, Reich PB, van Gestel N, et al. The economic value of grassland species for carbon storage. Science Advances. 2017;**3**:e1601880. DOI: 10.1126/sciadv.1601880

[8] Leblanc HA, Russo RO. Carbon sequestration in an oil palm crop system (Elaeis guineensis) in the Caribbean lowlands of Costa Rica. Proceedings of the Florida State Horticultural Society. 2008;**121**:52-54

[9] Mohammed AM, Robinson JS, Midmore D, Verhoef A. Carbon storage in Ghanaian cocoa ecosystems. Carbon Balance and Management. 2016;**11**(6). DOI: 10.1186/s13021-016-0045-x

[10] Kaghazchi T, Soleimani M. Effect of raw materials on properties of activated carbons. Chemical Engineering and Technology. 2006;**29**:1247-1251. DOI: 10.1002/ceat.200500298

Chapter 2

Soil Carbon Sequestration through Agronomic Management Practices

Sikander Khan Tanveer, Xingli Lu, Shamim-Ul-Sibtain Shah, Imtiaz Hussain and Muhammad Sohail

Abstract

Improper soil and crop management practices have resulted in loss of soil carbon. Worldwide, about 1417 Pg of soil carbon is stored in first meter soil depth, while 456-Pg soil carbon is stored in above–below ground vegetation and dead organic matter. Healthy soils can be helpful in combating the climate change because soils having high organic matter can have higher CO_2 sequestration potential. Main agronomic practices responsible for soil carbon loss include improper tillage operations, crop rotations, residue management, fertilization, and similarly no or less use of organic fertilizers that have resulted in the loss of soil organic matter in the form of CO_2. The share of agriculture sector in the entire emissions of global GHGs in the form of CO_2, N_2O, and CH_4 is about 25–30%. Studies have shown that by adapting proper tillage operations, the use of such kind of crop rotations that can improve soil organic matter and similarly the application of organic fertilizers, i.e., FYM, compost, and other organic amendments such as humic acid, vermicompost, etc., can be useful in soil carbon sequestration.

Keywords: soil carbon, agronomic practices, tillage, crop rotation, crop residues, organic fertilizers

1. Introduction

Soil carbon (C) sequestration implies the removal of atmospheric CO_2, by plants and storage of the fixed C through incorporation into soil organic matter [1]. Carbon exists in a variety of forms, mainly as plant biomass, soil organic matter, and gas carbon dioxide (CO_2) in atmosphere and dissolved in sea water. Soil organic carbon (SOC), which is a main component of SOM, can be separated into stable and labile fraction [2], and soil organic matter and its contribution play a very vital role during its humification formation of stable humus fraction and in the management of fertilization [3]. Worldwide, about 1417 Pg of soil carbon is stored in first meter soil depth, while 456-Pg soil carbon is stored in above–below ground vegetation and dead organic matter. The Earth's soils include approximately 1500 Pg of C, which is about 2–3 times larger than the amount of C stored in Earth's vegetation [4, 5]. The atmospheric carbon pool contains ~800 Pg of CO_2-C and is escalating at the rate of 4.2 Pg C per year, 0.54 percent per year. Over the past 150 years, the amount of carbon in the atmosphere has enlarged by 30%. An increase in the atmospheric concentration of CO_2 from 280 ppm from the pre-industrial era to 390 ppm in 2010 (an enrichment of 39%) and other greenhouse gases (GHGs) has changed the

Earth's mean temperature and precipitation [6]. There is much interaction among the terrestrial and atmospheric C pools through the processes of photosynthesis and respiration. Due to land use, conversion factors, and deforestation, biotic pool also contributes in the rise of atmospheric CO_2 concentration at the rate of ~1.6 Pg C per year. Different anthropogenic sources include the combustion of fossil fuel, deforestation, land use conversion, soil tillage, animal husbandry, cement manufacturing, etc. According to an estimate, 8.3 Pg C $year^{-1}$ is emitted by combustion of fossil fuel [6, 7], and 1.6 Pg C per year is emitted by deforestation, land-use change, and soil cultivation. It is anticipated that terrestrial ecosystems have contributed as much as half of increases in CO_2 emissions from human activity in the past two centuries [4, 8], and about 50 Pg CO_2 additions to the atmosphere has been contributed by cultivated soils [9], through the process of mineralization of soil organic carbon (SOC). Terrestrial C pool is estimated approximately 3120 Pg, which is the combination of both pedologic and biotic C pools.

Historically, agricultural soils have lost more than 50 Gt (1 Gt = 1 billion tons) of carbon and agriculture is responsible for soil carbon reductions up to 60–75% [9]. Total anthropogenic emission of CO_2 is 9.9 Pg C per year, of which 4.2 Pg C per year is absorbed by atmosphere and 2.3 Pg C per year by the ocean while remaining may be absorbed by unidentified terrestrial sinks.

In 1-m soil depth, estimated carbon pool is 2500 Pg, in two diverse forms including soil organic C (SOC) pool which is likely about 1550 Pg and soil inorganic C (SIC) pool at 950 Pg [10]. Soil inorganic C pool mostly consists of elemental C and carbonate minerals, i.e., calcite, dolomite, and likewise primary and secondary carbonates, whereas soil organic C (SOC) pool contains highly active humus and relatively inert charcoal C. According to United Nations Framework Convention on Climate Change (UNFCCC), carbon sequestration is the process of removing C from atmosphere and depositing it in a reservoir. It entails the transfer of atmospheric CO_2 and its secure storage in long-lived pools [11].

The estimation of global carbon sequestration potential of agricultural soils is typically made for sequestration on annual basis, and its range is from 0.4 to 1.2 gigatons per year [1]. Land use, land use change, and forestry (LULUCF) activities can be a relatively cost-effective ways to offset emissions through increasing removals of greenhouse gases from the atmosphere (e.g., by planting trees or managing forests) or through dropping emissions (e.g., by deforestation) [12]. Likewise, emissions of CO_2 from soil can be reduced by the adoption of such practices that can increase C input in soils and similarly can lessen the decomposition potential of soil organic matter. These kinds of practices have a vital role in storage and in release of C within terrestrial C cycle [13]. Nowadays, intensive agriculture usually results in a considerable soil degradation and soil carbon depletion [14], because in present agriculture and human's food chain, intensive soil utilization is very essential but it is very imperative so it should be followed and coupled with appropriate conservation practices [15]. Agriculture sector is responsible for the emissions of about 30% global greenhouse gases emissions, and primarily, inappropriate soil and crop management practices have resulted in the loss of soil carbon. In agricultural soils, C sequestration means the increase of soil C storage.

Main agronomic and related practices that can be helpful in SOC sequestration include:

- adoption of no-tillage (NT) or minimum tillage;
- adoption of environmental and soil health friendly farming systems;
- incorporation of cover crops;

- use of mulch either in the form of crop residues or synthetic materials;
- minimization of soil and water losses by surface runoff and erosion;
- adoption of integrated nutrient management practices for the increase of soil fertility;
- use of organic amendments; and
- promotion of farm forestry.

Benefits of soil carbon sequestration include the following:

- It can be helpful in the reduction of CO_2 emissions.
- It can reduce the emissions of different GHGs.
- It can be helpful in the reduction of atmospheric temperatures.
- It helps in maintaining suitable biotic habitat.
- It decreases nutrients losses.
- It can improve soil health and productivity.
- It can increase water conservation.
- It can promote and sustain root growth.
- It can reduce soil erosion.

Agriculture sector can be supportive in the lessening of emissions of GHGs, and if suitable agronomic practices are to be adopted, then agricultural soils have the potential to act as a sink for CO_2 sequestration. Healthy soils can be supportive in combating the climate change because soils having high organic matter can have higher CO_2 sequestration potential.

2. Agronomic practices

Different agronomic and related practices that can be supportive in CO_2 sequestration are given below.

2.1 Tillage

The main aim of tillage is the physical disturbance of upper soil layers for the preparation of soil bed, incorporation of fertilizers, crop residues, and similarly to control weeds. Tillage methods in world vary depending upon the soil, climate, crop management, and availability of technology. The relationship between tillage, soil structure, and soil organic matter dynamics is essential to C sequestration ability of agricultural soils. Tillage effects on soil carbon dynamics are complex and often variable [16]. Global reductions in natural SOC due to cultivation by humans are obvious, and it is estimated to cause a loss of 60 (temperate regions) to 75%

(temporal regions) of the original SOC [17]. Conventional tillage practices led to decline in soil carbon from 30 to 50% globally [18] to low as 20% [19]. Plowing is the basic cause of SOC oxidation and emissions of CO_2 to the atmosphere [20], and when NT, CP, and MP are under a nonsteady state, all these types of tillage systems may fail in the sequestration of significant amount of soil organic carbon [21]. The large losses of C typically follow initial cultivation [22, 23]. Moldboard plow, followed by secondary tillage operations, is commonly used in world, which is basically intensive tillage practice, but over the several years, intensive tillage has replaced by less intensive tillage in which soil is minimum disturbed. No tillage often increases the stability and numbers of soil aggregates, but conventional tillage is detrimental to soil structure, which increases the decomposition of soil organic matter. Conservation tillage systems keep more crop residues on the soil surface and have a higher SOC concentration in surface layer than conventional tillage [24, 25].

Tillage and cropping systems can influence microbial activity, which ultimately affects SOC dynamics and stability [26, 27], and soil mineralization can be decreased by reducing or eliminating soil tillage and increasing cropping intensity and plant production efficiency. In case of no-tillage as litter accumulates at the soil surface, which reduces evaporation from the soil because surface residues [28] and similarly standing stubbles [29] decrease wind speed at the soil surface, which ultimately results in less turbulent exchange of water and heat. Reduction in soil temperature through the use of surface mulches and no-till practices is important for maintaining stocks of soil organic matter especially in tropical soils [30].

SOC is a prime determinant of biological activity and soil macro fauna, which controls most of the different soil functions, i.e., organic matter dynamics, nutrient release, soil structure, and its different associated physical properties [31, 32]. In no-tilled soils, there are generally higher densities of biota and especially microorganisms. A large number of studies have shown that no-tillage can increase soil carbon rapidly, particularly at the soil surface [33], and this increase is linked to increases in aggregation [34, 35]. Compared to the PT and RT systems, strong SOC gradients have been observed under NT systems in the surface to subsurface layers in paddy soil. Moreover, it has been observed that the impacts of tillage on SOC concentration are dependent on crop species and soil depth in paddy soil [36]. However, according to Grandy and Robertson [37], tilling a previously untilled soil quickly losses the previously reserved carbon gains by exposing carbon molecules to microbial attack due to the disruption of aggregates. This accelerated turnover also reduces the formation and stabilization of more recalcitrant organic matter fractions within micro aggregates that have a longer residence time in soil [38]. The results of a study, which was conducted to find out the influence of conservation tillage, land configuration, and residue management practices on soil health in a Pigeon pea+ Soybean intercropping system. The study consisted of six tillage systems, i.e., CT_1: conservation tillage with BBF and crop residue retained on the surface, CT_2: conservation tillage with BBF and the incorporation of crop residue, CT_3: conservation tillage with flatbed with crop residue retained on the surface, CT_4: conservation tillage with the incorporation of crop residue, CT_5: conventional tillage with the incorporation of crop residue, and CT_6: conventional tillage without crop residue. The conservation treatments significantly improved soil health. The pooled data of the study showed that all the conservation tillage systems, i.e., CT_1, CT_2, CT_3, and CT_4, had significantly higher soil organic carbon at 0–15 cm depth (0.62, 0.64, 0.60, and 0.62%, respectively) and at 15–30 cm depth (0.56, 0.56, 0.54, and 0.55%, respectively) in higher soil carbon sequestrations (15.07, 15.39, 14.58, and 14.72 t ha^{-1}, respectively), over conventional systems. The study also revealed that however biological soil quality, such as soil microbial biomass carbon and nitrogen, was significantly higher in all the tillage systems except conventional tillage without crop residue [39]. It is estimated that the adoption of conservation tillage globally can sequester 25 Gt C over the next 50 years, which can be helpful in the stabilization of atmospheric carbon [40].

All this indicates that the adoption of conservation tillage practices can be helpful in the reductions of emissions of CO_2 into the atmosphere and similarly can be supportive in the sequestration of carbon in the soil.

2.2 Nutrient management

Chemical fertilizers are a source of emission of GHGs, especially N_2O. In addition to it, fertilizer production and its transportation are also associated with the emissions of GHGs. Judicious use of fertilizers increases crop yields and profitability, and about 50 Pg CO_2 additions to the atmosphere has been contributed by the cultivated soils [9], through the process of mineralization of soil organic carbon (SOC). The use of fertilizers has dramatically increased agricultural productivity, but studies reveal that the chronic use of nitrogen fertilization decreases soil microbial activity [41–44]. Continuous use of balanced fertilizers is necessary for sustainable soil fertility and productivity of crops [45]. Crop residues and nutrients, especially N, help in carbon sequestration up to 21.3–32.5% [46]. However, ultimate effects of continuous nitrogen fertilization on soils are complicated and remain unclear. For example, in the long-term experiments in Canada, SOC sequestration were 50–75 g cm^{-2} per year in well-fertilized soils with optimum cropping systems [47]. Research in the Great Plains shows that SOC sequestration is improved by the application of N fertilization [48–52], but opposite to it, long-term experiments in the Northern Great Plains (ND) have also shown that N fertilizer increased crop residue returns but generally did not increase SOC sequestration [53]. Liu Enke et al. [54] reported the results of a long-term study which was initiated in Northwest China in 1979, to find out the effects of fertilization on SOC and SOC fractions for the whole soil profile such as (0–100 cm) soil depth. The experiment included six treatments, i.e., unfertilized (control), N fertilizer (N), nitrogen and phosphorous fertilizer (NP), straw plus N and P fertilizers (NP + S), Farmyard manure (FYM), and Farmyard manure (FYM) plus N and P fertilizers (NP+ FYM). Results showed that SOC storage in 0–60 cm in NP + FYM, NP + S, FYM, and NP treatments increased by 41.5, 32.9, 28.1, and 17.9%, respectively, as compared to control treatment. Application of organic manure plus inorganic fertilizer also enlarged labile pool in 0–60 cm soil depth. These results show that long-term applications of organic manure have the most beneficial effects in building carbon pools among the investigated types of fertilization.

The results of Morrow plots, which is the world's oldest experimental site under continuous corn (*Zea mays* L.), revealed that after 40–50 years of synthetic fertilization that exceeded grain N removal by 60–190%, a net decline occurred in soil C despite increasingly massive residue C incorporation, the decline being more extensive for a corn-soybean (*Glycine max* L.) or corn-oats (*Avena sativa* L.) rotations than for the continuous corn rotation and of greater intensity for the profile (0–46 cm) than the surface soil [55]. Nayak et al. (2012) [56] reported that the application of combined inorganic fertilizers with or without manure can sequester carbon in the 0–60 cm soil layer at the Indean Sub-Himalayas. Majumder et al. [57] reported the results of a study that was conducted in hot humid subtropical Eastern India. According to them after 19 years in a puddle rice-wheat (*Triticum aestivum* L.) system, NPK + FYM treated plots had 14% larger labile C pools compared with the control plots in the 0–60 cm soil depth.

It can be concluded that the appropriate use of fertilizers according to the soil condition can be helpful in the maximum sequestration of carbon along with maximum crops production and in the reductions of emissions of different GHGs.

2.3 Animal manure and compost application

Animal manure is animal's excreta which is collected from livestock farms and barnyards and is used to enrich the soil, while compost is the material which largely consists of decayed organic matter and is used for fertilizing and conditioning of agricultural soil. Application of manures is important for the maintenance of soil health [58, 59] and is the source of C, and its application to different crops fields has effects on C contents [60]. As compared with the application of only NPK, application of FYM along with NPK increased C sequestration in the rice-wheat cropping system [61], while green manuring, as compared with the application of FYM along with green manure, sequestered more C in a Maize-Wheat cropping system [62]. Composting not only increases the net primary production but also enhances the C contents of the soil [63]. It has been reported that decreasing of manures and organic fertilizers application influences not only stable organic compounds but also soil microorganisms and nutrient regimes [64, 65]. Liu et al. [53] supported the positive effect of incorporation of mineral fertilizers with organic manures. Similarly, application of different organic wastes, i.e., municipal solid waste (MSW), farm yard manure (FYM), sugar industry waste (filter cake), and maize cropping residues, at 3 t C ha^{-1} alone and with a full or half dose of NPK mineral fertilizer showed that the addition of organic wastes (filter cake or MSW) has the best potential for improving SOC retention, WUE, and wheat yield in an irrigated maize-wheat cropping system [66].

This all indicates that the use of animal manure, compost, etc. along with other inorganic fertilizers is beneficial for both soil health and environment.

2.4 Crop rotations

Crop rotations mean the sequence of crops grown in regularly recurring successions on the same area of land. The succeeding crops may be for 2 or more years. Differences in crop rotations, climates, soils, and different crop-related management practices also affect carbon sequestration. Intensive cropping systems result in the depletion of SOM, but the use of balanced fertilization with NPK, application of organic amendments, and similarly application of crop residues can increase carbon sequestration levels to 5–10 Mg ha^{-1} per year because these amendments contain 10.7–18% C, which can also be helpful in the sequestration of carbon [67]. Different legume crops, such as peas, lentils, alfalfa, chickpea, sesbania, etc., can serve as substitute sources for nitrogen. Applications of crop rotations especially by using legume cover crops, which contain carbon compounds that are likely more resistant to microbial metabolism, can make soil carbon more stable [68]. Syswerda et al. [69] reported the results of a long-term study (over a 12-year period) of an organic management system that involved various crop rotations. According to them despite of extensive tillage for weed control, increase in soil carbon sequestration was recorded. The results of a long-term study, which was conducted in Dingxi, Northwest China, during 2013–2015, were shown in-spring wheat-field pea rotation in a rain-fed semi-arid environment. The treatments were: conventional tillage with stubble removed (T); no tillage with stubble removed (NT); no-till with stubble retained (NTS), and conventional tillage with stubble incorporation (TS). The SOC, microbial biomass carbon, and root biomass in NTS increased over T and NT, and similarly, average grain yield across the 3 years in NTS was better than T and

NT [70]. Recently, much attention has been given to alternate tillage and cropping systems as a means to mitigate the agricultural emissions of CO_2 [27, 71]. Different types of cropping systems, i.e., cover cropping, ratoon cropping, and companion cropping, can be helpful in carbon sequestration. Intercropping which includes row inter cropping, strip inter cropping, mixed cropping, and relay intercropping can increase the income and can also raise soil fertility [72]. Some of the examples of inter cropping are wheat and mustard, cotton and peanut, peanut and sunflower, wheat and chickpea, etc. [73]. Organic farming can also improve soil organic carbon as compared with the conventional farming [68, 74]. Research regarding the restoration of grassland also shows that through their biotic and biotic effects, legume species have more positive effects on the restoration of grasslands as compared with the application of mineral fertilizers [75].

This above shows that keeping in view the economic considerations, selection of appropriate crop rotations according to the soil and environmental conditions can be helpful in the sequestration of carbon, which not only improve soil fertility but also reduce the emissions of CO_2 into the atmosphere and increase farmer's income.

2.5 Residues management

Crop residues are detached vegetative parts of crop plants that are intentionally left to decay in agricultural fields after crop harvesting. Worldwide, the annual production of crop residues is about 3.4×10^9 tones, and if 15% of these total residues are applied to the soil, it can increase the C contents of the soil, because, for example, one ton of cereal residue contains 12–20 kg N, 1–4 kg P, 7–30 kg K, 4–8 kg Ca, and 2–4 kg Mg. Mulching is detached vegetation, which includes wheat straw, compost, or may be plastic sheets, which are spread around plants to protect them from excessive evaporation and cold stress and similarly to promote SOM contents in soil.

Crop residues play an important role in the SOC management and improvement of soil quality [76]. Mulching improves soil moisture, reduces soil erosion, and similarly reduces the loss of carbon from the soil and crop residues, which are incorporated into the soil to enhance the soil organic matter. A direct seedling mulch-based cropping system increases soil organic matter, as a result of increased carbon inputs and decreased soil disturbance [27]. Mulch can increase soil organic matter (SOM) and carbon sequestration in the top 0–5 cm soil depth. It improves soil's physical and chemical properties and can increase carbon sequestration in agricultural soils up to 8–16 Mg ha^{-1} per year. Mulch-based cropping systems enhance the buildup of soil organic matter, principally as a result of increased carbon inputs and decreased soil disturbance [27]. Direct seedling straw mulch has the potential to ameliorate the heat stress, and it improves the infiltration rate, reduces evaporation [77, 78], and similarly increases soil organic carbon and N efficiency [79]. Increasing residues inputs to soils entails increasing net primary productivity (NPP). Many agricultural soils, which have been significantly reduced from their original C levels through cultivation, will show C gains in proportion to increases in C inputs. Soil C levels are governed by the balance between the inputs of C through plant residues and the losses of C basically through decomposition. Therefore, C can be increased in soil by increasing residues inputs and or reducing decomposition rates (i.e., heterotrophic soil respiration). Litter quality also affects rates of its decomposition [80]. The results of a 4-month study, which was conducted in a greenhouse controlled condition and in three rates of straw residue and farm yard manure, were added to uncultivated and cropland soils. Two treatments of straw residue and farm yard manure incorporation were used into: a soil surface layer and a 0–20 cm soil depth revealed that the application of organic matter, especially the incorporation of farm yard manure, led to significant increase in the final soil organic carbon content, and higher amount of soil organic carbon were stored in the cropland soil than in the uncultivated soil. The results showed that carbon sequestration ranged farm yard manure > straw residue and cropland soil > uncultivated soil. The results revealed paying more attention to the role of organic residue management in carbon sequestration [81].

This all shows that the application of mulch and the use of crop residues can improve soil microbial activity, ameliorate the heat stress, and help in water storage and improvement of soil organic carbon.

2.6 Cover crops

Cover crop is grown for the benefit of soil rather than the crop yield. Cover crops improve soil quality by increasing soil organic carbon through biomass, by improving soil aggregates and stability, and by protecting the soil from surface runoff. Similarly, green manuring increases the biomass returned to the soil, which results in the form of enlarged soil carbon sink. Studies reveal that the adoption of cover crops is an efficient measure to mitigate climate change [82]. According to Olson 2010 [83], the use of cover crops in intensive row crop rotations with different tillage treatments has been found to sequester soil organic carbon (SOC). Kenneth et al. [84] reported the results of a study which included different kinds of tillages, i.e., no-till (NT), Chisel plow (CP), and moldboard plow (MP) with and without cover crops. The average annual corn and soybean yields were statistically same with or without cover crops. The average annual corn and soybean yields were statistically same for NT, CP, and MP systems with or without cover crops for the same soil depth layer and for tillage treatments. However, all tillage treatments, i.e., NT, CP, and MP, sequestered SOC with cover crops.

Keeping in view the cropping systems, suitable selection and planting of cover crops can be helpful in improving the soil organic carbon.

2.7 Use of improved crop varieties

Selection of improved varieties of different crops, which can improve both above and below ground biomass, can also improve the soil organic carbon. Machado et al. [85] reported that crop species that have massive rooting systems have the potential to improve SOC in soils under NT. Similarly, according to Kell [86, 87] by improving root growth in agricultural crops, soil carbon storage can match anthropogenic emissions for the next 40 years. This all indicates that the use of improved crop varieties having extensive root systems and better yields can increase both yields and soil fertility.

2.8 Soil biota management

Soil microbial activities can be helpful in the biological carbon sequestration because microbes improve the soil physical, chemical, and biological properties. The soil biota consists of a large number and a range of micro- and macro-organisms and is the living part of soils. They interact with each other and with plants, directly providing nutrition and other benefits. Their physical structure and products help a large to soil structure. They are also responsible for organic matter decomposition and for the transformations of organically bound nitrogen and minerals that are available to plants. Through biological control mechanisms, these organisms regulate their own populations and as well as those of incoming microorganisms. Micro- and macro-organisms are very crucial in maintaining ecosystem function, and their populations are significantly affected by the different crop management practices. Microorganisms include bacteria, fungi, fungi, protozoa, and some nematodes. These also include a range of invertebrates such as micro- and macro-arthropods, termites, and earthworms. According to an estimate, micro-organisms constitute about one quarter of the total biomass on the Earth [88]. These organisms are affected by the management of soils in the agricultural and forest ecosystems. Soils also differ in their ability to support the survival and growth of different groups of micro- and macro-organisms. Research findings show that carbon sequestration was higher up to 49.9 g C kg^{-1} in soils which were rich in soil microbes such as soil bacteria and fungi [89]. Therefore, the use of different kinds of microbes, which are beneficial both for soil and environment, will increase soil carbon sequestration and improve the crops yields.

2.9 Bio char

Bio char is carbonized biomass, which is obtained from sustainable sources and sequestered in soils. It can also be obtained by pyrolysis synthetically. Application of Bio char can also improve the soil health through carbon sequestration, because it improves the crop yield and maintains the cation exchange capacity, water holding, and nutrient retention capacity of the soil. It remains stable for thousands of years and thus reduces the release of terrestrial C to the atmosphere in the form of CO_2 [90]. It has been reported that Bio char can improve carbon sequestration in soil due to prolonged residence time [91]. Another study also reveals that the application of Bio char reduces the co-localization of polysaccharides-carbon and aromatic carbon by reducing the carbon metabolism due to carbon stabilization in Bio char-activated soil [92]. It has also been reported that soil management by using different kinds of organic amendments and their incorporation by earthworms can also support micro-aggregates formation, C, and N retention in agricultural soils [93].

2.10 Agroforestry

Agroforestry is the combination of agriculture and forestry in which perennial trees and shrubs are grown in combination with agricultural crops. Planting of different kinds of trees, including orchards, fruit plants, and woodlands into the croplands, can improve soil carbon sequestration. Agroforestry has an enormous potential for carbon sequestration in croplands [94] because agroforestry practices accumulate more C than forests and pastures because they have both forest and grassland sequestration and storage patterns active [95–97]. Young [98] have also reported the estimated potential of C gains from agroforestry. Agricultural soils can sequester more quantities of carbon by the adoption of agroforestry. The carbon sequestrations potential of agroforestry systems is estimated between 12 and 228 Mg ha^{-1}, so on the Earth's total suitable area for crop production, a total of about 1.1–2.2 Pg C can be sequestered in the agricultural soils in the next 50 years [99]. The results of a meta-analysis from 53 published studies, regarding changes in soil organic carbon (SOC) stocks at 0–15, 0–30, 0–60, 0–100, and $0 \geq 100$ cm, after land conversion to agroforestry, revealed a significant decline in the SOC stocks of 26 and 24% in land-use changes from forest to agroforestry at 0–15 and 0–30 cm, respectively. The transition from agriculture to agroforestry significantly enhanced the SOC stock of 26, 40, and 34% at 0–15, 0–30, and 0–100 cm, respectively. The results also showed that conversion from pasture/grassland to agroforestry produced significant SOC stock increases at 0–30 cm (9%) and 0–30 cm (10%). Switching from uncultivated/other land-uses to agroforestry increased SOC by 25% at 0–30 cm, while a decrease was observed at 0–60 cm (23%) [100].

The carbon sequestration potential by agroforestry is estimated up to 9, 21, 50, and 63 Mg Cha^{-1} in semiarid, subhumid, humid, and temperate regions, respectively; however, it has been reported that intensively managed agroforestry practice in combination with annual crops is like conventional agriculture, which does not contribute in carbon sequestration [101].

Agroforestry also helps in the optimization of water use, and similarly, it improves the farmer's income. So, the promotion of agroforestry keeping in view the soil condition, climate, and along with crops production is beneficial for soil, environment, as well as the farmers.

3. Conclusion

CO_2 is increasing at the rate of 2.3 ppm per year, which is resulting in the increase of global warming and environmental pollution. Agriculture sector is responsible for up to 30% emission of GHGs. Sustainable agriculture is essential for the survival of humankind. Adoption of different agronomic management practices can be helpful in the sequestration of carbon. Such practices include no-tillage or reduced tillage, nutrient management, cover crops, crop rotations, green manuring, application of animal manures, agroforestry, etc. Adoption of these different agronomic practices will not only improve the crops yields but will also improve farmer's income.

Author details

Sikander Khan Tanveer[1]*, Xingli Lu[2], Shamim-Ul-Sibtain Shah[3], Imtiaz Hussain[1] and Muhammad Sohail[4]

1 Wheat Program, Crop Sciences (CSI), PARC–National Agricultural Research Center (NARC), Islamabad, Pakistan

2 College of Agronomy Ningxia Agriculture University, P.R China

3 Directorate of Farm Operations and Services, National Agricultural Research Center (NARC), Islamabad, Pakistan

4 Plant Physiology Program, Crop Sciences (CSI), PARC–National Agricultural Research Center (NARC), Islamabad, Pakistan

*Address all correspondence to: sikander73@hotmail.com

References

[1] Lal R. Soil carbon sequestration to mitigate climate change. Geoderma. 2004;**123**:1-22

[2] Sun Y, Huangs YUX, Zhang W. Stability and saturation of soil organic carbon in rice fields: Evidence from a long-term fertilization experiment in subtropical China. Journal of Soils and Sediments. 2013;**13**(8):1307-1334. DOI: 10.1007/s 11368-013-0741

[3] Yang R, Y-z S, Wang T, Yang Q. Effect of chemical and organic fertilization on soil carbon and nitrogen accumulation in a newly cultivated farmland. Journal of Integrative Agriculture. 2016;**15**(3):658-666. DOI: 10.1016/S 2095-3119(15)61107-8

[4] Post WM, Peng T-H, Emanuel WR, King AW, Dale VH, De Angelis DL. The global carbon cycles. American Scientist. 1990;**78**:310-326

[5] Eswaran H, Van Den Berg E, Reich PF. Organic carbon in the soils of the world. Soil Science Society of America Journal. 1993;**57**:192-194

[6] IPCC. Climate Change 2007. The Fourth Assessment Report. The Physical Science Basis. Cambridge, United Kingdom: Cambridge University Press; 2007

[7] WMO. Greenhouse Gas Bulletin: The State of the Art Greenhouse Gases in the Atmosphere until December. Geneva: World Meteorological Organization; 2010. p. 2009

[8] Houghton RA, Skole DL. Carbon. In: Turner BL, Clark WC, Kates RW, Richards JF, Mathews JT, Meyer WB, editors. The Earth as Transformed by Human Action. Cambridge UK: Cambridge University Press; 1990. pp. 393-408

[9] Lal R. Sequestering carbon in soils of agro-ecosystems. Food Policy. 2011;**36**:S33-S39

[10] Batjes NH. Total C and N in soils of the world. European Journal of Soil Science. 1996;**47**:151-163

[11] UNFCCC. Report of the Conference of Parties on its Thirteenth Session, Bali, Indonesia. Geneva: United Nations Framework Convention on Climate Change; 2007

[12] UNFCCC.2012. Land Use, Land Use Change and Forestry (LULUCF). Available from: http:// Unfcccint/ methods_and_Science/lulucf/ items/3060.php

[13] Lal R, Kimble JK, Levin E, Stewart BA, editors. Advances in Soil Science: Soil Management and Greenhouse Effect. Boca Ratoon, FL: Lewis Publishers; 1995. p. 93

[14] Plaza- Bonilla D, José Luis A, Cantero-Martinez C, Fanlo R, Iglesias A, Alvaro-Fuentes J. Carbon management in dryland agricultural systems. A review. Agronomy for Sustainable Development. 2015;**35**(4):1319-1334. DOI: 10.10007/s 13593-015-0326-x

[15] Lal R. Enhancing crop yields in the developing counteries through restoration of soil organic carbon pool in agricultural lands. Land Degradation and Development. 2006;**2**(17):197-209. DOI: 10.1002/Idr-696

[16] Sombrero A, de Benito A. Carbon accumulation in soil. Ten-year study of conservation tillage and crop rotation in semi-arid area of castle- Leonion Spain. Soil and Tillage Research. 2010;**107**:64-70

[17] Lal R. Soil carbon sequestration impacts on global climate change

and food security. Science. 2004;**304**(5677):1623-1627

[18] Schlesinger WH. Changes in soil carbon storage and associated properties with disturbance and recovery. In: Trabalha JR, Reichle DE, editors. The changing carboncycle: A global analysis. New York: Springer-Verlag; 1985. pp. 194-220

[19] Sharma PK, Bhushan L, Ladha JK, Naresh RK, Gupta RK, Balasubramanian V, et al. Crop - water relations in rice-wheat systems and water management practices in a marginally sodic, medium textured soil. In: Bouman et al., editors. Water-wise rice production. LOS Banos (Philippines): International Rice Research Institute; 2000. pp. 223-235

[20] Al-Kaisi MM, Yin X. Tillage and crop residue effects on soil carbon and carbon dioxide emission in corn-soybean rotations. Journal of Environmental Quality. 2005a;**34**: 437-445. DOI: 10.2134/jeq2005.0437

[21] Olson KR. Impacts of tillage, slope and erosion on soil organic carbon retention. Soil Science. 2010;**175**:562-567. DOI: 10.1097/SS.09o13e3181foc 2837.

[22] Davidson EA, Ackerman IL. Changes in soil carbon inventories following cultivation of previous untilled soils. Biogeochemistry. 1993;**20**:161-164

[23] Haas HJ, Evans CE, Miles EF. Nitrogen and carbon changes in Great Plains soils as influenced by cropping and soil treatments. Technical Bulletin No.1164 USDA, State Agricultural Experiment Stations; 1957

[24] Drury CF, Tan CS, Welacky TW, Oloya TO, Hamil AS, Weaver SE. Red clover and tillage influence on soil temperature, water content, and corn emergence. Agronomy Journal. 1999;**19**:101-108

[25] Huchinson JJ, Campbell CA, Desjardins RL. Some perspectives on carbon sequesteration in agriculture. Agricultural Meteorology. 2007;**142**:288-307

[26] Scow KM. Soil microbial communities and carbon flow in agro ecosystems. In: Jackson LT, editor. Ecology in agriculture. San Diego, CA: Academic press; 1997. pp. 367-413

[27] Paustian K, Collins HP, Paul EA. Management controls on soil carbon. In: Paul EA, Paustin K, Elliott ET, Cole CV, editors. Soil Organic Matter in Temperate Agroecosystems. Boca Raton, and FL: CRC Press; 1997. pp. 15-49

[28] Bond JJ, Willis WO. Soil water evaporation: Surface residue rate and placement effects. Soil Science Society of America Proceedings. 1969;**33**:445-448

[29] Smika DE. Soil water change as related to position of wheat straw mulchon soil surface. Soil Science Society of America Journal. 1983;**47**:988-991

[30] Lal R. Conservation tillage for sustainable agriculture: Tropic Vs. temperate environments. Advances in Agronomy. 1989;**42**:84-191

[31] Lavelle P, Spain AV. Soil Ecology. Dordrecht, The Netherlands: Kluwer Academic Publishers; 2001. p. 654

[32] Blanchart E, Albrecht A, Brown GG, Dacaëns T, Duboisset A, Lavelle P, et al. Effects of tropical endogenic earthworms on soil erosion: A review. Agriculture, Ecosystems and Environment. 2004;**104**:303-315

[33] West TO, Post WM. Soil organic carbon by tillage and crop rotation: A global data analysis. Soil Science Society of America Journal. 2002;**66**(6):1930-1946

[34] Cambardella CA, Elliott ET. Particulate soil organic –matter changes across a grassland cultivation sequence. Soil Science Society of America Journal. 1992;**56**(3):777-783

[35] Six J, Elliott ET, Paustian K. Soil macro aggregate turnover and micro aggregate formation: A mechanism for C sequestration under no-tillage agriculture. Soil Biology and Biochemistry. 2000;**32**(14):2099-2103

[36] Xu SQ, Zhang MY, Zhang HL, Chen F, Yang GL, Xiang XP. Soil organic carbon stocks as affected by tillage systems in a double-cropped rice field. Pedosphere. 2013;**23**:696-704

[37] Grandy AS, Roberson GP, Thelen KD. Do productivity and environmental trade-offs justify periodically cultivating no-till cropping systems? Agronomy Journal. 2006;**98**(6):1377

[38] Six J, Elliott ET, Paustian K. Aggregate and SOM dynamics under conventional and no tillage systems. Soil Science Society of America Journal. 1999;**63**:1350-1358

[39] Naveen Kumar BT, Babalad HB. Soil organic carbon, carbon sequestration, soil microbial biomass carbon and nitrogen and soil enzymatic activity as influenced by conservation agriculture in pigeonpea and soybean intercropping system. International Journal of Current Microbiology and Applied Sciences. 2018;**7**(03):323-333. DOI: 10.20546/ijcmas.2018.703.038

[40] Pakala S, Socolow R. Stabilization wedges: Solving the climate problem for the next 50 years with current technologies. Science. 2004;**305**:968-972

[41] Marschner P, Kandeler E, Maschner B. Structure and function of the soil microbial community in a long - term fertilization experiment. Soil Biology and Biochemistry. 2003;**35**(3):453-461

[42] Bowden RD, Davidson E, Savage K, Arabia C, Steudler P. Chronic nitrogen additions reduce total soil respiration and microbial respiration on temperate forest soils at the Harvard Forest. Forest Ecology and Management. 2004;**196**(1):43-56

[43] Ramirez KS, Craine JM, Fierer N. Consistent effects of nitrogen amendments on soil microbial communities and processes across biomes. Global Change Biology. 2012;**18**(6):1918-1927

[44] Frey SD, Ollinger S, Nadelhoffer K, Bowden R, Brzostek E, Burton A, et al. Chronic nitrogen addition suppress decomposition and sequester soil carbon in temperate forests. Biogeochemistry. 2014;**121**(2):305-316

[45] Verma G, Sharma RP, Sharma SP, et al. Changes in soil fertility of maize – wheat system due to long-term use of chemical fertilizers and amendments in an alfisol. Plant, Soil and Environment. 2012;**58**:529-533

[46] Windeatt JH, Ross AB, Williams PT, Forster PM, Nahil MA, Singh S. Characteristics of Bs from crop residues: Potential for C sequestration and soil amendment. Journal of Environmental Management. 2014;**146**:189-197

[47] Domanski J, Desjardins RL, Tarncai C, Monreal D, Gregorich EG, Kirkwood V, et al. Possibility for future carbon sequesteration in Canadian agriculture in relation to land use changes. Climatic Change. 1998;**40**:81-103

[48] Campbell CA, Zentner RP. Soil organic matter as influenced by crop rotations and fertilization. Soil Science Society of America Journal. 1993;**57**:1034-1040

[49] Campbell CA, Zentner RP, Liang B-C, Rooloff G, Gregorich EC,

Blomert B. Organic C accumulation in soil over 30 years in semi-arid southern western Saskatchewan- effect of crop rotations and fertilizers. Canadian Journal of Soil Science. 2000a;**80**:179-192

[50] Halvorson AD, Reule CA, Follet RF. Nitrogen fertilization effects on soil carbon and nitrogen in a dryland cropping systems. Soil Science Society of America Journal. 1999c;**63**:912-917

[51] Halvorson AD, A-Reule C, Murphy LS. No-tillage and N fertilizer enhance soil carbon sequestration. Journal of Fluid. 2000c;**8**:8-11

[52] Nyborg M, Solberg ED, Malhi SS, Izaaurralde RC. Fertilizer N, crop residue, and tillage after soil C and N contents after a decade. In: Lal R et al., editors. Advances in Soil Science: Soil Management and Greenhouse Effect. Boca Raton FL: Lewis Publishers, CRC Press; 1995. pp. 93-100

[53] Halvorson AD, Weinhold BJ, Black AL. Tillage, nitrogen and cropping systems effects on soil carbon sequestration. Soil Science Society of America Journal. 2002;**66**:906-912

[54] Liu E, Yan C, Mei X, Zhang Y, Fan T. Long-term effect of manure and fertilizer on soil organic carbon pools in dryland farming in Northwest China. PloS One. 2013;**8**(2):e56536. DOI: 10.1371/Journal pone.0056536

[55] Khan S, Mulvaney RL, Ellsworth T, Boast CW. The myth of nitrogen fertilization for soil carbon sequestration. Journal of Environmental Quality. 2007;**36**(6):1821-1832

[56] Nayak AK, Gangwar B, Shukla AK, Muzumdar SP, etal KA. Long-term effect of different integrated nutrient management on soil organic carbon and its fractions on soil organic carbon and its fractions and substainability of rice-wheat systems in indo Gangetic Plains of India. Field Crops Research. 2012;**127**:129-139

[57] Majumder B, Mandal B, Bandyopadhyay PK, Gangopadhyay A, Mani PK, et al. Organic amendments influence soil organic carbon pools and rice-wheat productivity. Soil Science Society of America Journal. 2008;**72**:775-785

[58] Schlesinger WH. Carbon sequestration in soils. Science. 1999;**284**:2095

[59] Baker JM, Ochsner TE, Venterea RT, Griffis TJ. Tillage and soil carbon sequestration – What do we really know? Agriculture, Ecosystems and Environment. 2007;**118**:1-5

[60] Stewart CE, Paustian K, Conant RT, Plante AF, Six J. Soil C sequestration: Concept, evidence and evaluation. Biogeochemistry. 2007;**86**:19-31

[61] Naresh RK, Gupta RK, Minhas PS, Rathore RS, Ashish D, Purushottam V. Climate change and challenges of water and food security for smallholder farmers of Uttar Pradesh and mitigation through C sequestration in agricultural lands: An overview. International Journal of Chemical Studies. 2017;**5**(2):221-236

[62] Kukal SS, Rasool R, Benbi DK. Soil organic C sequestration in relation to organic and inorganic fertilization in rice-wheat and maize-wheat systems. Soil and Tillage Research. 2009;**102**:87-92

[63] Baldi E, Cavani L, Margon A, Quartieri M, Sorrenti G, Marzadori C, et al. Effect of compost application on the dynamics of C in a nectarine orchard ecosystem. Science of the Total Environment. 2018;**162**:239-248

[64] Zhang W, Xu M, Wang X, Huang Q, Nie J, Li Z, et al. Effects of organic amendments on soil carbon

sequestration in paddy fields of subtropical China. Journal of Soils and Sediments. 2012;**12**(4):457-470. DOI: 10.1007/s11368-011-0467-8

[65] Ren T, Wang J, Chen Q, et al. The effects of manure and nitrogen fertilizer application on soil organic carbon and nitrogen in a high input cropping system. PloS One. 2014;**9**(5):e97732. DOI: 10.1371/journal.pone 0097732

[66] Shehzadi S, Shah Z, Mohammad W. Impact of organic amendments on soil carbon sequestration, water use efficiency and yield of irrigated wheat. Biotechnology, Agronomy, Society and Environment. 2017;**21**(1):36-49

[67] Manadal B et al. The potential of cropping systems and soil amendments for carbon sequestration in soils under long-term experiments in subtropical India. Global Change Biology;**13**:357-369

[68] Wickings K, Grandy AS, Read SC, Cleveland CC. The origin of litter chemical complexity during decomposition (Njohnson Ed.). Ecology Letters. 2012;**15**(10):1180-1188

[69] Syswerda SP, Corbin AT, Mokma DL, Kravchenko AN, Roberson GP. Agricultural management and soil carbon storage in surfaces vs deep layers. Soil Science Society of America Journal. 2011;**75**(1):92

[70] Yeboah S, Zhang R, Cai L, Li L, Xie J, Lou Z, et al. Tillage effect on soil organic carbon, microbial biomass carbon and crop yield in spring wheat – field pea rotation. Plant, Soil and Environment. 2016;**62**(6):279-285

[71] Follet RF. Soil management concepts and carbon sequesteration in cropland soils. Soil and Tillage Research. 2001;**61**:77-92

[72] Hirts KKK. Mixed Cropping, Agricultural Techniques Known as Mixed Cropping. 2009. Available from: http://archaeology.About.com/od/history of agriculture/qt/mixed cropping.Html [Accessed: 28 March 2010]

[73] Anon, "Mixed Cropping". Available from: http://Simple.wikipedia.Org/wiki/Mixed_Cropping.Html

[74] Kowatsuzali M, Syuaib MF. Comparison of the farming systems and carbon sequestration between conventional and organic rice production in West Java, Indonesia. Sustainability. 2010;**2**:833-843

[75] De Deyn GB, Shiel RS, Ostle NJ, McNamara NP, Oakley S, Young I, et al. Additional carbon sequestration benefit, of grassland diversity restoration. Journal of Applied Ecology. 2011;**48**:600-608

[76] Chen Zhang-du, Zhang Hai-lin, Dikgwatlhe S-Batsile, Xue Jian-fu, Qiu Kang-Chang, Tang H ai-ming, et al. Soil carbon storage and stratification under different tillage/residue – Management practices in double rice cropping system. Journal of Integrative Agriculture. 2015:1-16. Advance on line publication. DOI: 10.1016/s2095-3119(15)61068-1

[77] Lal R. Role of Mulching Techniques in Tropical Soil and Water Management. Ibadan, Nigeria: Tech. Bull ITTA; 1975

[78] Lal R. Tillage in lowland rice-based cropping systems. In: Soil Physics and Rice. Philippines: IRRI; 1985. pp. 283-307

[79] Hobbs PR, Gupta RK. Sustainable resource management intensively cultivated irrigated rice-wheat cropping systems of the Indo-Gangetic Plains of South Asia. Strategies and options. In: Singh AK, editor. Proceedings of the International Conference on Managing Resources For Sustainable Agricultural Production. In the 21st century. New Delhi, India 14-18 February 2000. New

Delhi, India: Indian Society of Soil Science; 2000. p. 157

[80] Heal OW, Anderson JM, Swift MJ. Plant litter quality and decomposition: An historical overview. In: Cadisch G, Giller KE, editors. Driven by Nature-Plant Litter Quality and Decomposition. Wallingford: CAB International; 1997. pp. 3-30

[81] Mahmoodabadi M, Heydarpour E. Sequestration of organic carbon influenced by the application of straw residue and farmyard manure in two different soils. International Agrophysics. 2014;**28**:169-176. DOI: 10.2478/intag-2014-0005

[82] Vicente - Vicente JL, Garcia-Ruiz R, Francaviglia R, Aguilera E, Smith P. Soil carbon sequestration rates under Mediterranean woody crops using recommended management practices: A meta-analysis. Agriculture Ecosystems and Environment. 2016;**235**:204-214

[83] Olson KR, Ebelhar SA, Lang JM. Cover crop effects on on crop yields and organic carbon content. Soil Science. 2010;**175**:89-98. DOI: 10.1097/ss.ob0133e3181cf7959

[84] Olson K, Stephen A, Ebelhar JML. Long-term effects of cover crops on crop yields, soil organic carbon stocks and sequestration. Open Journal of Soil Science. 2014;**4**:284-292

[85] Machado S, Rhinhart K, Petrie S. Longterm cropping systems effects on carbon sequestration in eastern Oregon. Journal of Environmental Quality. 2006;**35**:1548-1553

[86] Kell DB. Breeding crop plants with deep roots: Their role in sustainable carbon, nutrient and water sequestration. Annals of Botany. 2011;**108**(93):407-418

[87] Kell DB. Large scale sequestration of atmospheric via plant roots in natural and agricultural ecosystems: Why and how. Philosophical Transactions of the Royal Society B. 2012;**367**(1595):1589-1597

[88] Jong SC. Microbial germplasm. In: Knutson L, Stone AK, editors. Biotic Diversity and Germplasm Preservation, Beltsville Symposia in Agricultural Research. Kluwer Academic Publications: Dordrecht; 1988, 1989. pp. 241-273

[89] Bailey VL, Smith JL, Bolton H. Fungal -to- bacterial ratios in soils investigated for enhanced C sequestration. Soil Biology and Biochemistry. 2002;**34**:997-1007

[90] Lehmann J. Bio-energy in the black. Frontiers in Ecology and the Environment. 2007;**5**:381-387

[91] Hemandez-Soriano MC et al. Long-term effect of bio char on the stabilization of recent carbon: Soils with historical inputs of charcoal. GCB Bioenergy. 2016;**8**(2):371-381

[92] Hemandez-Soriano MC et al. Bio char affects carbon composition and stability in soil: A combined spectroscopy-microscopy study. Scientific Reports. 2016;**6**:25127

[93] Pulleman MM, Six J, Uyl A, Marinissen JCY, Jongmans AG. Earthworms and management affect organic matter incorportation and microaggregates formation in agricultural soils. Applied Soil Ecology. 2005;**29**:1-15

[94] Sanches PA, Buresh RJ, Leakey RRB. Trees, soils and food security. Phillosphical Transactions of the Royal Society of London, Series B. 1999;**353**:949-961

[95] Schroeder P. Carbon storage benefits of agroforestry systems. Agroforestry Systems. 1994;**27**:89-97

[96] Kort J, Turnock R. Carbon reservoir and biomass in Canadian prairie shelterbelts. Agroforestry Systems. 1999;**44**:175-186

[97] Sharrow SH, Ismail S. Carbon and nitrogen storage in agroforests, tree plantations, and pastures in Western Oregon, USA. Aggroforest Systems. 2004;**60**:123-130

[98] Young A. Agroforestry for Soil Management. 2nd ed. CABI; 1 December 1997. pp. 257-258

[99] Albrecht A, kandji ST. Carbon sequestration in tropical agroforestry systems. Agriculture Ecosystems and Environment. 2003;**99**:15-27

[100] De Stefano A, Jacobson MG. Soil carbon sequestration in agroforestry systems: A meta-analysis. Agroforestry Systems. 2018;**92**:285-299. DOI: 10.1007/s 10457-017-0147-9

[101] Montagnini F, Nair PKR. Carbon sequestration: An under exploited environmental benefit of agroforestry systems. In: Nair PKR, Rao MR, Buck LE, editors. New Vistas in Agroforestry: A Compendium for 1st World Congress of Agroforestry. Dordrecht: Springer Netherlands; 2004. pp. 281-295

Chapter 3

Landuse and Physiographic Region Effects on Soil Carbon and Nitrogen Sequestration in Arkansas

Marya McKee, Kristofor R. Brye and Lisa Wood

Abstract

Increasing understanding of soil carbon (C) sequestration dynamics and general functioning in disappearing native grassland ecosystems, has the potential to enhance soil rehabilitation and ecosystem restoration. The objective of this study was to evaluate the effects of landuse (native tallgrass prairie and managed agriculture) and physiographic region (northwest Arkansas and east-central Arkansas) on the change in soil C and nitrogen (N) storage and other soil properties over a 15-year period. Despite the native prairie losing soil C at a rate of 4.7 Mg ha^{-1} $year^{-1}$ over the 15-year duration of this study, soil C storage in 2016 was more than 2.5 times greater in the native prairie than in the cultivated agroecosystems in the Grand Prairie. Averaged across landuse, TC concentration ($P < 0.01$) and content ($P < 0.01$) changed more over time in the Ozark Highlands region of northwest Arkansas (0.02% $year^{-1}$ and 0.28 Mg ha^{-1} $year^{-1}$, respectively), than in the Grand Prairie region of east-central Arkansas. This study demonstrates the value of direct measurements over time for assessing temporal changes in soil properties and results can potentially direct future restoration activities to be as successful as possible.

Keywords: carbon sequestration, silt-loam soils, managed grassland, cultivated cropland, and native prairie

1. Introduction

Greenhouse gas (GHG) concentrations in the atmosphere have been on the rise since the Industrial Revolution began in the eighteenth century and have led to an enhancement of the greenhouse effect and dramatic increases in global air temperatures. The combined average land and surface air temperatures from 1880 to 2012, calculated from a linear trend with 90% certainty, showed mean warming of 0.85°C, which ranged from 0.65 to 1.06°C, across the globe [1]. This striking increase in air temperature over such a short period of time has evolved into the climatic variations witnessed today that scientists have termed global climate change. The scientific consensus is that the rising air temperatures are the result of an unprecedented rise in GHG emissions over the last 150 years, due primarily to human activity in the form of burning fossil fuels, agricultural land conversion and management, and deforestation [1].

Carbon dioxide (CO_2) is the most abundant GHG in the atmosphere and has risen from 280 mg L^{-1} before the Industrial Revolution in 1730 to over 400 mg L^{-1} in 2015 and is predicted to continue increasing by 2 mg L^{-1} per year [1]. In 2010, the CO_2 derived from burning fossil fuels and industrial processes contributed around 65% of total global GHG emissions, while forestry, land clearing for agriculture, and soil degradation collectively accounted for 11% of the global emissions [1]. In 2015, agricultural production alone accounted for 9% of total US GHG emissions [2]. Fossil fuels are burned during agricultural production, from heavy machinery and equipment, and these emissions account for nearly 25% of the total global GHG emissions originating from fossil fuels [1]. However, other sources of CO_2 emissions associated with agricultural production result from the initial conversion and continued tillage of soils that contain substantial amounts of soil organic carbon (SOC).

Carbon flows through the global C cycle between five primary pools: the oceanic pool, geologic pool (fossil fuels), pedologic (soil) pool, atmospheric pool, and biotic pool. Most of the C held within the terrestrial ecosystem is stored within the soil (~2500 Pg), with only a fraction of the total terrestrial C stored within plant biomass (~560 Pg) [3]. The global C cycle and SOC creation are fueled by the photosynthetic activity of plants and other autotrophic organisms, which convert CO_2 into glucose ($C_6H_{12}O_6$) or energy that is used to develop biomass. Following the decomposition of autotrophic organisms and the heterotrophic organisms that consume them, a portion of the C captured during photosynthesis remains stored in the soil in a process known as soil C sequestration.

Soil C sequestration potential within an ecosystem is dependent on the same factors that lead to pedogenesis: parent material, climate, biota, topography, and time. Climate and biota tend to play a more significant role in the C cycle and therefore are often more important to the soil C sequestration process. Colder climates are more likely to accumulate SOC because soil microbial decomposition of soil organic matter (SOM) is slowed and sometimes paused when temperatures reach below freezing (0°C) [4]. In addition to temperature, the other main climatic factor that affects soil C sequestration is moisture, where a fine line exists between optimal soil moisture and too much or too little soil moisture. Optimal soil moisture contributes to increase in above- and belowground autotrophic productivity, which leads to greater available plant biomass. Too much soil moisture decreases soil microbial decomposition rates of SOC when soil moisture levels reach a point where periodic reducing conditions occur. However, increased moisture levels that do not reach reducing conditions have the potential to increase soil microbial decomposition rates, as microbes gain more access to the soil pores through the movement of water. In contrast, too little soil moisture limits plant productivity, hence the potential production and input of SOM for soil microbes to consume and convert to SOC.

The C compounds returned to the soil during microbial decomposition, in the form of SOM, consist of three separate pools: active, intermediate or slow, and passive. These pools are classified by the amount of time the SOM remains stable before further decomposing and returning to the atmosphere, mostly as CO_2, via microbial respiration. Soil organic matter within the active pool consists of labile or easily decomposable particulate organic matter [5]. The active pool of SOM contributes most of the beneficial effects of soil structural stability, which enhances a soil's infiltration capacity and resistance to erosion and can be readily increased by adding fresh plant and animal residues. However, due to the potential instability of the organic material, the active C pool can be easily lost due to reductions in organic additions or increased tillage.

Agricultural production is one of the leading disturbances of the SOC pools because of the management practice of conventional tillage [6]. Over the years, due to reduced C inputs and continual disturbance through conventional tillage, between 40 and 60% of SOC has been lost following the conversion of lands from tallgrass prairie to cultivated agriculture [7]. The decrease can be explained by the initial land conversion to non-native vegetation, followed by the rapid oxidation of SOM from conventional tillage and minimal vegetative cover and biomass inputs back to the soil. Rapid declines in SOC are due partly to the mechanical disintegration of soil aggregates during repeated annual tillage, which exposes organo-mineral surfaces that had otherwise been unavailable for decomposition by microbes when previously undisturbed and the physical loss of SOC due to erosion [8]. However, some level of SOM turnover is necessary to incorporate and protect fresh C inputs from rapid decay and mineralization, whereas SOM turnover that occurs too quickly or too often can lead to decreases in microaggregate stability and formation, therefore decreasing soil C sequestration potential [9].

Soil formation and resulting soil properties are inextricably linked with climatic factors and parent material. Arkansas resides in a unique climatic transition zone in the mid-southern United States, with arid to semi-arid grassland to the west and northwest and humid-temperate deciduous forest to the south and southeast. The climate gradient that spans through Arkansas also controls the botanical composition and interacts with distinct topographic and geologic gradients. Consequently, Arkansas is separated into at least two distinct climates and parent material sources. The northwest region of Arkansas, known as the Ozark Highlands [Major Land Resource Area (MLRA) 116A] [10], is relatively warm and wet and dominated by deciduous forest vegetation both presently and as the climax vegetation community. Cherty limestone residuum is the soil parent material for much of the Ozark Highlands. Soils in the Ozark Highlands are typically moderately deep and medium- to fine-textured Udults and Udalfs. The Ozark Highlands also contains remnants of the Osage Prairie, which once extended through south-central and southwestern Missouri, as well as northwest Arkansas [11]. Presently, less than 0.5% of the original Osage prairie exists, due to the conversion to pasture and hay meadows now populated with naturalized (i.e., introduced) species [11]. East-central Arkansas contains the remnants of the tallgrass prairie regionally referred to as the Grand Prairie, within MLRA 134—Southern Mississippi Valley Silty Uplands [10], which is an area that is also relatively warm and wet, with fertile alluvial parent material.

Comparing native prairies across these two physiographic regions in Arkansas, Brye et al. [12] reported that SOM concentrations in the top 10 cm in native prairie ecosystems generally increased south and eastward across Arkansas, due to increasing precipitation in the Grand Prairie region compared to the Ozark Highlands, therefore leading to increase above- and belowground productivity. In contrast, a study conducted by Brye and Gbur [13] reported that, averaged across landuse, soils in the Grand Prairie region had the lowest average SOC sequestration rate (−0.04 kg SOC m^{-2} $year^{-1}$) in the top 10 cm compared to that observed in the Ozark Highlands (0.05 kg SOC m^{-2} $year^{-1}$). Brye and Gbur [13] also reported that, averaged across physiographic region, soils under agricultural management had a lower average SOC sequestration rate (−0.03 kg SOC m^{-2} $year^{-1}$) in the top 10 cm compared to that observed for native prairies (0.04 kg SOC m^{-2} $year^{-1}$).

A study conducted within the Ozark Highlands region evaluating the effects of grassland management on soil physical and chemical properties, including SOC, in the top 10 cm over a 6-year period in silt-loam soil reported that certain management schemes, including grazing and haying, had minimal effect on near-surface soil properties, indicating that contemporary managed forage lands in the Ozark Highlands are not being degraded by agricultural management, but are in fact remaining similar to the remnant prairies from which they were converted [14]. In contrast, a study conducted, within the Grand Prairie region, on a chronosequence of four Typic Albaqualfs in adjacent fields in Prairie County, Arkansas, varying only in time under cultivation with the control as a native, undisturbed tallgrass prairie, reported an exponential decrease in SOC in the top 10 cm as years of cultivation increased [15]. Brye and Pirani [16] showed a significant difference in soil C concentration and content in the top 10 cm between native prairie (2.3–3.2% C; 2.5–3.4 kg C m^{-2}) and adjacent tilled agricultural systems (1.0–1.7% C, 1.3–2.0 kg C m^{-2}) in the Grand Prairie region of east-central Arkansas. Specifically, the difference in soil C tended to be greater when the agricultural fields had been tilled for 30 or more years than when the agricultural fields had been tilled for less than 30 years [16]. Following the conversion from native prairie to intensely tilled agriculture, a 17–52% decrease in soil-quality-related parameters was observed, including soil OM, total N, and total C [16].

Many of these examples of losses in SOC, and other resulting soil-quality parameters, can be interpreted positively as the potential to enhance and regain soil C storage in soils affected by long-term agricultural activity. With a combination of site-specific best management practices, like conservation tillage, no-tillage, cover crops, and elimination of fallow periods, which increase C input into the system, agricultural fields have the potential to become C sinks instead of their historic role as C sources. The objective of this study was to evaluate the effects of landuse (i.e., native tallgrass prairie and managed agriculture) and physiographic region on the change in soil C storage and other soil properties over a 15-year period. It was hypothesized that soil C and N storage in the top 10 cm would remain constant or slightly increased within the prairie remnants, while soil under agricultural management (i.e., managed pastureland, and cultivated agriculture) would likely decrease over a 15-year period between 2001 and 2016. Cultivated row-crop agriculture on alluvial soils was hypothesized to have lower C and N storage than managed pastureland on residual soils due to the long history of intensive tillage, despite alluvial soils typically being considered more fertile than residual soils.

2. Materials and methods

2.1 Site description

The Stump and Chesney Prairies in Benton County in the Ozark Highlands region of northwest Arkansas have various degrees of disturbed, agricultural landuse adjacent to the prairies in the same soil map unit as exists in the prairies (**Table 1**). Specifically, adjacent to the Stump Prairie resides a managed grassland, dominantly tall fescue (*Lolium arundinaceum* [Schreb.] Darbys.), that has been infrequently cultivated in the last 20 years for cutting and removing the aboveground vegetation (i.e., haying) multiple times a year. In addition, a managed pastureland that has never been cultivated, but has been consistently grazed multiple times per year for the past 20 years with varying head of cattle, also resides adjacent to the Stump Prairie. Adjacent to the Chesney Prairie resides an area that has been periodically cultivated and planted with a row crop [i.e., corn (*Zea mays* L.) or soybean (*Glycine max* L.)] in the last 20 years; however, the area has been left fallow for the past 15 years without a row crop being planted.

The Seidenstricker Prairie, in Prairie County, Arkansas, resides in an area known as the Grand Prairie and has three adjacent agricultural fields, with similar soil mapping units as exists in the prairie, that were once part of the prairie itself

Region/parent material	Site	Landuse	Years managed	Soil series	Soil taxonomic description	Slope (%)
Ozark Highlands/ residuum	Stump	Periodically cultivated hayfield	>20	Jay	Oxyaquic Fragiudalf	0
	Stump	Never cultivated managed pasture	>20	Jay	Oxyaquic Fragiudalf	0
	Stump	Native prairie	0	Jay	Oxyaquic Fragiudalf	1
	Chesney	Periodically cultivated agriculture	>20	Jay	Oxyaquic Fragiudalf	2
	Chesney	Native prairie	0	Jay	Oxyaquic Fragiudalf	2
Grand Prairie/ alluvium	Seidenstricker	Cultivated agriculture	31*	Dewitt	Typic Albaqualf	0
	Seidenstricker	Cultivated agriculture	42*	Dewitt	Typic Albaqualf	0
	Seidenstricker	Cultivated agriculture	60*	Dewitt	Typic Albaqualf	0
	Seidenstricker	Native prairie	0	Dewitt	Typic Albaqualf	0

**Indicates years before 2016, therefore native prairie was converted to cultivated agriculture in 1986, 1975, and 1957, respectively.*

Table 1.
Summary of site characteristics by geographic region.

that have now been consistently annually intensively cultivated and cropped under a rice (*Oryza sativa* L.)-soybean-wheat (*Triticum aestivum* L.) rotation in most years since 1957, 1975, and 1986 (**Table 1**). Consequently, as of 2016, the native prairie and three adjacent agricultural fields represented a chronosequence with varying years under cultivated agriculture (i.e., 0, 31, 42, and 60 years, respectively).

2.2 Regional characteristics

The Ozark Highlands (36–38°N lat., 91–95°W long.), MLRA 116A [10], occupies portions of southwest and south-central Missouri, eastern Oklahoma, and northwest and north-central Arkansas. The area is a low-elevation, disjointed mountainous region, covering roughly 2.1 million ha [13]. Soils in the Ozark Highlands are typically Udults and Udalfs with deep, medium- to fine-textured cherty residuum weathered from limestone [17]. Oak (*Quercus* spp.) forests dominate the vegetation in the Ozark Highlands, but a large extent of tallgrass prairie was also historically present. The Chesney and Stump Prairies are located within the Springfield Plateau, a region that extends over 640,000 ha, which consisted of low, undulating plains (240–430 m in elevation) covered in prairie, savannah, hardwood forest, and acidic glade ecosystems [10]. Historic prairie ecosystems covered >30,000 ha within the Springfield Plateau. The Chesney and Stump prairies are the only remnants of a much grander historical prairie that spanned over 4000 ha in northwest Arkansas known as the Lindsley Prairie [18].

The Grand Prairie (34°0′–35°30′ N lat., 91°15′–92°10′ W long.), part of MLRA 134, is in east-central Arkansas and covers roughly 0.5 million ha [10]. Soils in the Grand Prairie are typically deep to very deep Udalfs, with medium texture and mixed mineralogy [17], that are developing in fertile alluvial parent material sourced from the historic flooding of the Mississippi River, with or without a thin loess cover. The historic land cover in the Grand Prairie region was grasslands, which historically covered ~130,000 ha as tallgrass prairie, of which <1% remains today due primarily to the introduction and expansion of mechanized agriculture [19]. Consequently, presently, the predominant landuse within the Grand Prairie region is cultivated, row-crop agriculture, where rice, soybean, and wheat are the dominant crops.

The regions also vary by climate. The climate of the Grand Prairie region is, on average, warmer than the Ozark Highlands, with mean annual temperatures of 16.6 and 14.5°C, respectively [20]. Average annual precipitation in the northwest Arkansas portion of the Ozark Highlands is approximately 116 cm, while that in the Grand Prairie region is 126 cm [20].

2.3 Soil sampling scheme

Between early August 2001 and mid-April 2002, initial soil samples were collected at Stump and Chesney Prairies in the Ozark Highlands region of northwest Arkansas and at the Seidenstricker Prairie in the Grand Prairie region of east-central Arkansas. Soil property data in the top 10 cm from the initial soil samples collected in 2001/2002 were determined and reported in Brye and West [11] and Brye et al. [21] for the Stump and Chesney prairies and in Brye and Slaton [22] and Brye et al. [12] for the Seidenstricker Prairie. At the same time, between early August 2001 and mid-April 2002, the above-described, adjacent agricultural areas were sampled at all three prairie sites. Between late October and early November 2016, a subsequent set of soil samples were collected in all three prairie sites and in their adjacent agricultural areas. In each soil map unit represented within each prairie and in the same or similar soil map unit in the adjacent agricultural areas, soil samples were collected from the top 10 cm along a 60-m transect at five sampling points spaced 15 m apart (i.e., at the 0-, 15-, 30-, 45-, and 60-m marks). A slide hammer, with a 4.8-cm-diameter, stainless steel core chamber, was used to manually collect the soil samples, which were subsequently oven-dried at 70°C for 48 h, weighed for bulk density determinations, and crushed and sieved to pass through a 2-mm mesh screen for soil chemical property determinations.

Percentages of sand, silt, and clay in the top 10 cm from the initial soil samples collected in 2001/2002 were determined and reported in Brye and West [11] and Brye et al. [21] for the Stump and Chesney prairies and in Brye and Slaton [22] and Brye et al. [12] for the Seidenstricker Prairie. Soil pH was potentiometrically measured using an electrode in a 1:2 (wt/vol) soil-to-water paste. Soil organic matter was determined by weight-loss-on-ignition after 2 h at 360°C. Total C and N were determined by high-temperature combustion (Elementar Variomax CN Analyzer, Elementar Americas, Inc., Mt. Laurel, NJ). No soil among sampled transects effervesced upon treatment with dilute hydrochloric acid, thus all measured soil C was assumed to be SOC. The C:N ratio and fractionation of C and N in the SOM were calculated for each sample using measured concentrations. In addition, for each soil sample, TC, TN, and SOM contents (kg ha^{-1}) were calculated from measured concentrations (g kg^{-1}), measured bulk densities, and the 10-cm depth interval. To calculate C and N sequestration rates, the 2001/2002 contents were then subtracted from 2016 contents and the differences were divided by the number of years between sampling (~15 years).

2.4 Statistical analyses

A two-factor analysis of variance (ANOVA) was conducted using SAS 9.4 (SAS Institute, Inc., Cary, NC), based on a completely random design, to evaluate the effects of physiographic region (i.e., Ozark Highlands and Grand Prairie), landuse (i.e., native prairie and managed agriculture), and their interaction on changes in soil bulk density, pH, EC, SOM, C, and N storage, C:N ratio, and C and N fractions of SOM in the top 10 cm over time. A second two-factor ANOVA was conducted using SAS to evaluate the effects of physiographic region, landuse, and their interaction on soil bulk density, pH, EC, SOM, C, and N storage, C:N ratio, and C and N fractions of SOM in the top 10 cm from the 2016 sampling only to assess the current state of soil property differences among treatments. In addition, a linear regression analysis was conducted in Minitab (version 13, Minitab, Inc., State College, PA) using the 2016-measured data only for the Grand Prairie sites to assess soil C storage trends over time under cultivation. For all statistical analyses, significance was judged at $P < 0.05$; thus, when appropriate, means were separated by least significant difference (LSD) at the 0.05 level.

3. Results and discussion

3.1 Soil property differences after 15 years

In 2016, after 15 years of consistent management or natural time progression, all measured soil properties, with the exception of bulk density, C:N ratio, and the C and N fractions of SOM, differed ($P < 0.05$) between regions within landuses (**Table 2**). Soil bulk density and the C and N fractions of SOM differed ($P < 0.03$) between physiographic regions and differed ($P < 0.04$) between landuses, while the soil C:N ratio was unaffected ($P > 0.05$) by region or landuse (**Table 2**), thus was similar and averaged 13.5 across all region-landuse combinations (**Table 3**). **Table 3** also summarizes the means and standard errors among ecosystem-landuse combinations for soil properties from the original 2001/2002 soil sampling.

Soil pH and EC were greatest ($P < 0.01$; **Table 2**) in the Grand Prairie region within the agricultural landuse (6.7 and 0.168 dS m^{-1} respectively), and lowest in the native prairie landuse in the same region (4.7 and 0.072 dS m^{-1} respectively), but only pH and EC in the agricultural landuse in the Grand Prairie differed from the other region-landuse combinations (**Table 3** and **Figure 1**). Soil pH and EC were likely more regulated in the more conventional agricultural landuses in the Grand Prairie from annual fertilizer additions and irrigation, respectively. Brye and Pirani [16] similarly concluded that soil pH and EC were generally greater under tilled agricultural than under native prairie landuse.

In 2016, after 15 years of consistent management, SOM concentration and content were both more than two-fold greater in the Ozark Highlands under both landuses and in the Grand Prairie under prairie landuse, which did not differ, than in the Grand Prairie under cultivated agricultural landuse (**Figure 2**). Similarly, TN concentration and content were lowest, by more than 50%, in the Grand Prairie under cultivated agricultural landuse, while TN concentration was greatest under

Soil property	Region	Landuse	Region × landuse
		P	
BD (g cm^{-3})	0.03	<0.01	0.83
pH	<0.01	<0.01	<0.01
EC (dS m^{-1})	0.16	<0.01	<0.01
SOM (%)	<0.01	<0.01	<0.01
SOM (Mg ha^{-1})	<0.01	<0.01	<0.01
TN (%)	<0.01	0.02	0.04
TN (Mg ha^{-1})	<0.01	0.10	<0.01
TC (%)	<0.01	< 0.01	0.03
TC (Mg ha^{-1})	<0.01	< 0.01	<0.01
C:N ratio	0.26	0.45	0.32
TN fraction of SOM (%)	<0.01	0.04	0.92
TC fraction of SOM (%)	<0.01	0.03	0.60

Table 2.
Analysis of variance summary of the effects of physiographic region (Ozark Highlands and Grand Prairie), landuse (managed agriculture and native prairie), and their interaction on the change in soil bulk density (BD), pH, electrical conductivity (EC), soil organic matter (SOM), total nitrogen (TN), and total carbon (TC) concentration and content, C:N ratio, and the C and N fractions of SOM in the top 10 cm over a 15-year period in silt-loam soils in Arkansas.

prairie landuse in both regions, which did not differ, and TN content was greatest under prairie landuse in the Grand Prairie (**Table 4** and **Figure 1**). Total N concentration and content were also greater under prairie than managed agricultural landuse in the Ozark Highlands (**Table 4** and **Figure 2**). Similar to TN, TC concentration and content were more than twofold greater under prairie landuse in both regions, which did not differ, than under cultivated agriculture in the Grand Prairie (**Table 4** and **Figure 2**). Total C concentration and content were also greater under prairie than agricultural landuse in the Ozark Highlands (**Figure 2**). Differences in TC and TN content among region-landuse combinations were likely the result of combined differences in TC and TN concentrations and bulk density, where, averaged across landuse, bulk density was 1.1 times greater in the Grand Prairie than in the Ozark Highlands and, averaged across region, was also 1.1 times greater under managed agricultural than native prairie landuse (**Table 3**). However, the differences in TC and TN concentrations alone (**Table 4** and **Figure 2**) clearly demonstrate that there are substantial differences in the net balance between above- and/or belowground C and N inputs and losses.

Averaged across landuse, TN and TC fractions of SOM were both 1.3 times greater in the Ozark Highlands than in the Grand Prairie (**Table 3**). Averaged across region, TN and TC fractions of SOM were both 1.1 times greater under managed agricultural than native prairie landuse (**Table 3**). The differences in TC and TN fractions of SOM between regions and between landuses indicate that, in the Ozark Highlands and in the managed agricultural landuse use in general, the SOM pool is less diverse with other soil nutrients than in the Grand Prairie and native prairie landuse.

In contrast to the results of this study, Brye and Gbur [13] reported greater SOM, TN, and TC contents under the native prairie in the Grand Prairie, citing warmer and wetter annual climatic conditions that would promote greater belowground root biomass and OM, N, and C inputs compared with the Ozark Highlands.

Treatment	Soil properties†								
	BD (gcm^{-3})	pH	EC ($dS\ m^{-1}$)	SOM ($Mg\ ha^{-1}$)	TN ($Mg\ ha^{-1}$)	TC ($Mg\ ha^{-1}$)	C:N	N/SOM (%)	C/SOM (%)
Region									
Grand Prairie (GP)	1.21	5.7	0.120	39.5	1.24	17.0	13.7	3.22	44.0
Ozark Highlands (OZH)	1.14	4.9	0.108	54.3	2.33	30.7	13.3	4.29	56.8
Landuse									
Agriculture (AG)	1.23a	5.8	0.146	40.0	1.67	21.8	13.4	3.97a	52.6a
Prairie (PR)	1.11b	4.8	0.082	53.8	1.91	25.9	13.6	3.54b	48.2b
Ecosystem × landuse									
GP-AG	1.27	6.7a	0.168a	26.5b	0.89d	12.2c	13.7	3.38	46.4
GP-PR	1.15	4.7b	0.072b	52.5a	1.59c	21.7b	13.7	3.05	41.7
OZH-AG	1.20	4.9b	0.125b	53.6a	2.44a	31.3a	13.1	4.55	58.9
OZH-PR	1.07	4.8b	0.092b	55.0a	2.23b	30.1a	13.6	4.03	54.7
Ecosystem × landuse combination means (± standard error) from the 2001/2002 sampling††									
GP-AG	1.15 (0.02)	5.9 (0.1)	0.098 (0.01)	32.3 (1.3)	1.37 (0.03)	15.1 (0.3)	11.0 (<0.1)	4.34 (0.2)	47.9 (2.3)
GP-PR	1.13 (0.01)	4.7 (<0.1)	0.072 (<0.01)	58.1 (1.8)	2.33 (0.07)	28.9 (0.8)	12.4 (0.2)	4.02 (0.1)	49.9 (1.2)
OZH-AG	0.13 (0.04)	5.1 (0.1)	0.104 (0.01)	54.0 (2.7)	2.68 (0.17)	26.6 (1.6)	10.0 (0.3)	4.94 (0.1)	49.0 (1.0)
OZH-PR	0.06 (0.02)	4.8 (<0.1)	0.072 (<0.01)	49.2 (2.1)	2.18 (0.11)	24.1 (1.4)	11.0 (0.2)	4.43 (0.1)	48.7 (1.0)

†*Different lower case letters for a soil property within a treatment category indicates a significant difference ($P < 0.05$).*
††*Data reproduced from Brye and West [11], Brye et al. [21], Brye and Slaton [22], and Brye et al. [12].*

Table 3.
Summary of mean soil property changes by treatment (i.e., physiographic region, landuse and their interaction) over a 15-year sampling period for soil bulk density (BD), pH, electrical conductivity (EC), soil organic matter (SOM), total nitrogen (TN), total carbon (TC), C:N ratio, and the N (N/SOM) and C (C/SOM) fractions of SOM in the top 10 cm in silt-loam soils in Arkansas.

Similarly, comparing soil properties among only native prairies, Brye et al. [12] concluded that SOM and SOC concentrations and C:N ratio were at least numerically greater in the Grand Prairie than in the Ozark Highlands. However, Brye et al. [12] also reported that, based on a significant linear relationship, both TN and TC increased with increasing SOM concentration faster in the Ozark Highlands than in the Grand Prairie. Brye and West [11] reported that neither TN nor TC concentration differed in the top 10 cm between landuses in 2001/2002 when comparing the same sites used for this study in the Ozark Highlands. However, TN and TC concentration data were obtained through high-temperature combustion, whereas SOM concentration, which is obtained by weight-loss on ignition at a lower temperature, was significantly greater in the grazed than in the ungrazed pasture or native prairie soils [11]. These results indicate that the proportion of SOC within SOM likely differs between management systems, in which the quality of substrate entering

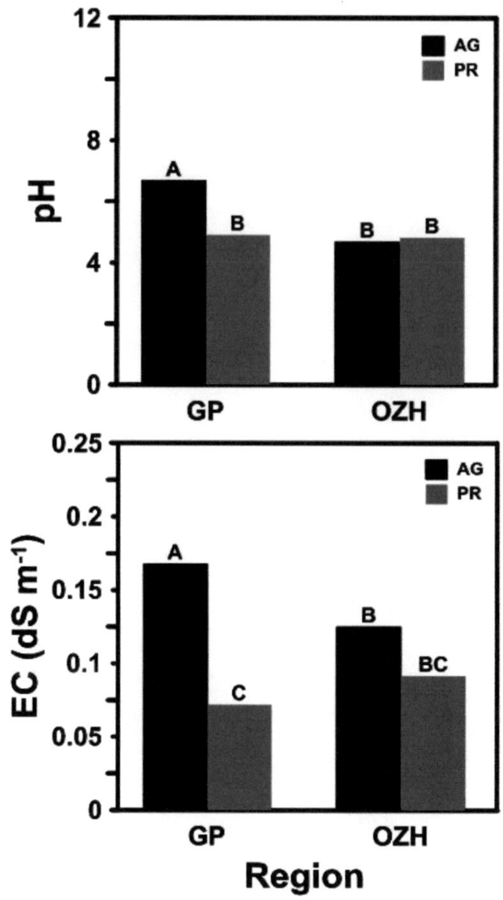

Figure 1.
Landuse effects by physiographic region on the change in soil pH and electrical conductivity (EC) in the top 10 cm over the 15-year sampling period under either agricultural management (AG) or undisturbed prairie (PR) landuse in the Grand Prairie (GP) region of east-central Arkansas or the Ozark Highlands (OZH) region of northwest Arkansas. Different letters associated with mean values on a panel are different at $P < 0.05$. An asterisk (*) indicates mean value is greater than 0 ($P < 0.05$).

the SOC pool through humification is likely more recalcitrant or physiochemically protected in the native prairie, indicating more C storage is occurring within the passive SOC pool.

3.2 Soil property changes over time

Most of the soil property differences measured in this study in the top cm over a period of 15 years from 2001/2002 to 2016 were affected by physiographic region, landuse, or both (**Table 5**). Changes in soil pH ($P < 0.01$) and EC ($P = 0.01$) over the 15-year period differed among regions within landuses (**Table 5**). In contrast, changes in SOM content, TC and TN content and concentration, C:N ratio, TC and TN fractions of SOM over time differed ($P < 0.05$; **Table 5**) between regions, while changes in TC content, C:N ratio, and TC fraction of SOM over time also differed ($P < 0.05$; **Table 5**) between landuses. Neither changes in soil bulk density nor SOM concentration over time were affected by region or landuse ($P > 0.05$; **Table 5**).

Soil pH and EC in the top 10 cm increased the most over time under cultivated, row-crop agricultural management in the Grand Prairie, which was a greater change over time than for the other three region-landuse combinations, which did

	Soil properties†		
Treatment	**SOM (%)**	**TN (%)**	**TC (%)**
Region			
Grand Prairie (GP)	3.33	0.12	1.44
Ozark Highlands (OZH)	4.88	0.21	2.75
Landuse			
Agriculture (AG)	3.31	0.14	1.81
Prairie (PR)	4.90	0.21	2.39
Region × landuse			
GP-AG	2.09b	0.07c	0.97c
GP-PR	4.58a	0.14b	1.92b
OZH-AG	4.53a	0.21a	2.64a
OZH-PR	5.23a	0.21a	2.86a

†*Different lower case letters for a soil property within a treatment category indicates a significant difference ($P < 0.05$).*

Table 4.
Summary of mean soil property changes by treatment (i.e., physiographic region, landuse, and their interaction) over a 15-year sampling period for soil organic matter (SOM), total nitrogen (TN), total carbon (TC) concentrations in the top 10 cm in silt-loams soils in Arkansas.

not differ (**Table 6** and **Figure 3**). However, soil pH and EC also decreased the most over time under non-cultivated agricultural landuse in the Ozark Highlands (**Table 6** and **Figure 3**). The differences in soil pH and EC change over time under agricultural landuse likely were due to periodic lime applications and exposure to bicarbonate-rich irrigation water for row-crop production in the Grand Prairie. In contrast, soil pH and EC did not change over time under native prairie landuse in either region, which may have occurred due to already having achieved some level of equilibrium that maintained the soil in a well-buffered state. Similar to the results of this study, based on samples collected in 1987 to a depth of 10 cm, also at the Seidenstricker site in the Grand Prairie, Brye et al. [15] concluded that soil pH levels were greater in the oldest cultivated, agriculturally managed soils, 12- and 30-year-old at the time, than in the native prairie and the youngest (1-year-old) cultivated, agriculturally managed soil. Following a resampling of the same sites in 2001, Brye et al. [15] reported that soil pH was still greater in the cultivated agroecosystems than in the prairie, but that soil pH did not differ among the three cultivated agroecosystems by 14 years later. Brye and Gbur [14], who compared soil property differences in the top 10 cm between native, undisturbed, and managed grasslands in the Ozark Highlands between 2001 and 2008, concluded that, although numerically greater in the agroecosystems, soil pH did not differ in the top 10 cm between native prairie and managed forage landuse, but soil pH levels had decreased by 8% in the 7 years between samplings. The soil pH decrease was attributed to a lack of liming in the managed forage lands and the presence of natural mineralization of the SOM and nitrification processes that had been slowly acidifying the soil [14].

In contrast to soil pH and EC, averaged across landuse, SOM content in the top 10 cm decreased over time in the Grand Prairie (-0.37 Mg ha^{-1} year^{-1}), but did not change over time in the Ozark Highlands (**Table 3**). Soil bulk density did not differ over time (**Table 5**), but, despite the change in SOM concentration over time being unaffected ($P > 0.05$) by region (**Table 5**), SOM concentration decreased

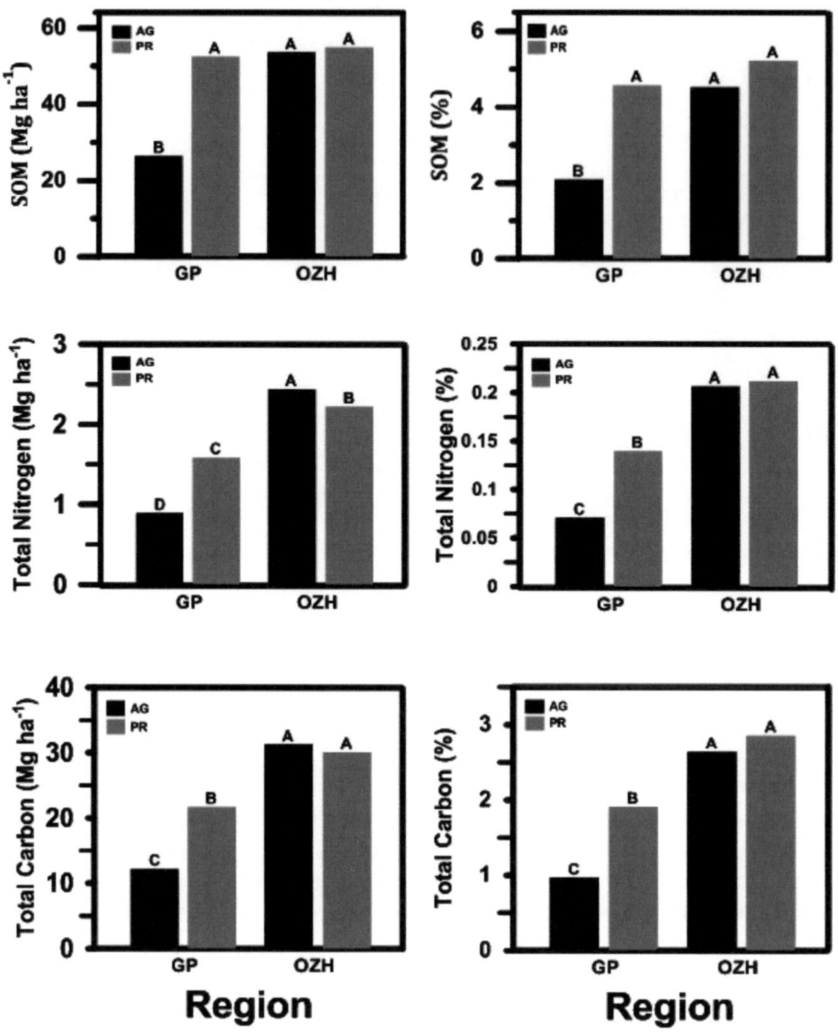

Figure 2.
Landuse effects by physiographic region on soil pH and electrical conductivity (EC) in the top 10 cm from the 2016 sampling only under either agricultural management (AG) or undisturbed prairie (PR) landuse in the Grand Prairie (GP) region of east-central Arkansas and the Ozark Highlands (OZH) region of northwest Arkansas. Different letters associated with mean values on a panel are different at P < 0.05.

more in the Grand Prairie than in the Ozark Highlands (**Table 6**), which was likely responsible for the decrease in SOM content in the Grand Prairie over time. Furthermore, the Grand Prairie region, on average, is slightly warmer and wetter than in the Ozark Highlands. Consequently, microbial decomposition of SOM was likely somewhat greater over time in the Grand Prairie than in the Ozark Highlands. Brye and Gbur [14] also concluded that SOM content did not change over time in either the native prairies or the managed grasslands in the Ozark Highlands region. In a study comparing landuse effects between the Ozark Highlands and the Grand Prairie regions among silt-loam-textured soils to a depth of 10 cm, between 2001 and 2007, Brye and Gbur [13] also demonstrated that, averaged across landuse, SOM content decreased in the Grand Prairie, but did not change over time in the Ozark Highlands and attributed the results to the regional climate differences. The Grand Prairie region also typically experiences longer durations of warmer temperatures, which stimulate microbial activity and lead to greater microbially mediated SOM decomposition rates.

Soil property	Region	Landuse	Region × landuse
		P	
BD (g cm^{-3} yr^{-1})	0.55	0.08	0.87
pH (yr^{-1})	<0.01	0.01	<0.01
EC (dS m^{-1} yr^{-1})	0.17	<0.01	0.01
SOM (% yr^{-1})	0.08	0.22	0.44
SOM (Mg ha^{-1} yr^{-1})	<0.01	0.59	0.68
TN (% yr^{-1})	0.05	0.76	0.15
TN (Mg ha^{-1} yr^{-1})	<0.01	0.57	0.16
TC (% yr^{-1})	<0.01	0.67	0.19
TC (Mg ha^{-1} yr^{-1})	<0.01	0.04	0.22
C:N ratio (yr^{-1})	0.04	<0.01	0.37
TN fraction of SOM (% yr^{-1})	0.02	0.99	0.96
TC fraction of SOM (% yr^{-1})	<0.01	0.06	0.76

Table 5.
Analysis of variance summary of the effects of physiographic region, landuse, and their interaction on soil bulk density (BD), pH, electrical conductivity (EC), soil organic matter (SOM), total nitrogen (TN), and total carbon (TC) concentration and content, C:N ratio, and the TN and TC fractions of SOM in the top 10 cm from the 2016 sampling in silt-loam soils in Arkansas.

Similar to SOM content, averaged across landuse, TN concentration (−0.004% $year^{-1}$; **Table 4**) and content (−0.04 Mg ha^{-1} $year^{-1}$; **Table 6**) decreased by more than a factor of two over time in the Grand Prairie respectively; table than in the Ozark Highlands, which also decreased over time (−0.002% $year^{-1}$ and −0.01 Mg ha^{-1} $year^{-1}$, respectively; **Tables 6** and 7). Despite the annual input of fertilizer N for optimal row-crop production in the Grand Prairie, which only meant to meet the crop N requirement, no other mechanisms of N input are enough to overcome the net loss of N over time from the top 10 cm by likely leaching and/or volatilization. In contrast to the results of this study, Brye and Gbur [13] reported that, averaged across landuses, TN content increased over time at a rate of 0.04 Mg ha^{-1} $year^{-1}$ in the Ozark Highlands, but decreased at the same rate (−0.04 Mg ha^{-1} $year^{-1}$) in the Grand Prairie in the top 10 cm between 2001 and 2007. However, in a chronosequence of tallgrass prairie restorations in the Ozark Highlands, TN content in the top 10 cm decreased over time with restoration age and trended toward that in a nearby native prairie, which contained the lowest TN content [23]. The variation in results among studies highlights the multifaceted and highly limiting nature of N in both natural and cultivated ecosystems.

Similar to TN, averaged across landuse, TC concentration (−0.03% $year^{-1}$; **Table 7**) and content (−0.33 Mg ha^{-1} $year^{-1}$; **Table 6**) decreased over time in the Grand Prairie. However, in contrast, TC concentration (0.02% $year^{-1}$, **Table 7**) and content (0.28 Mg ha^{-1} $year^{-1}$, **Table 6**) both increased over time in the Ozark Highlands. There are many complex processes that have been linked to fluctuations and accumulations of C within soils, such as plant physiological responses to atmospheric CO_2; light; and environmental stressors, like temperature, nutrients, and water, as well as microbial responses to soil moisture and temperature variations. In the case of this study, all of these factors are potentially at work, with likely a stronger influence stemming from microbial responses to differences in soil moisture and temperature between physiographic regions, where greater microbial decomposition is occurring in the slightly warmer and wetter climate of the Grand Prairie than in the Ozark Highlands. Erosion from wind and water can also play a major role in SOC

Treatment	Soil properties†								
	BD	pH	EC	SOM	TN	TC	C:N	N/SOM	C/SOM
Region									
Grand Prairie (GP)	0.005	0.027	0.002	−0.37b*	−0.04b*	−0.33b*	0.13b*	0.25	−0.32b*
Ozark Highlands (OZH)	0.003	−0.006	0.001	0.04a	−0.01a*	0.28a*	0.18a*	−0.04	0.52a*
Landuse									
Agriculture (AG)	0.006	0.019	0.003	−0.21	−0.02	0.06a	0.19a*	0.44	0.29
Prairie (PR)	0.001	0.001	0.000	−0.13	−0.03	−0.12b	0.12b*	0.05	−0.09
Region × landuse									
GP-AG	0.008	0.053a*††	0.005a*	−0.38	−0.03	−0.19	0.18	−0.09	−0.10
GP-PR	0.001	0.000b	−0.000b	−0.36	−0.05	−0.47	0.08	0.42	−0.54
OZH-AG	0.005	−0.015b*	0.001b*	−0.03	−0.02	0.32	0.21	0.03	0.68
OZH-PR	0.000	0.003b	0.000b	0.11	−0.01	0.24	0.16	−0.06	0.35

*An asterisk indicates mean value is greater than 0 ($P < 0.05$).
†Units for the soil properties are as follows: BD, g cm^{-3} yr^{-1}; pH, yr^{-1}; EC, dS m^{-1} yr^{-1}; SOM, TN, and TC, Mg ha^{-1} yr^{-1}; C:N, yr^{-1}; and N/SOM and C/SOM, % yr^{-1}.
††Different lower case letters for a soil property within a treatment category indicate a significant difference ($P < 0.05$).*

Table 6.
Summary of mean soil properties in the top 10 cm by treatment (i.e., physiographic region, landuse and their interaction) for soil bulk density (BD), pH, electrical conductivity (EC), soil organic matter (SOM), total nitrogen (TN), total carbon (TC), C:N ratio, and the N and C fractions of SOM from the 2016 sampling in silt-loam soils in Arkansas.

loss in soils, as well as the oxidation of SOC associated with cultivation, which could also be contributing to the overall loss of soil C in the Grand Prairie. Conventional tillage is carried out on an annual basis in the agricultural sites from the Grand Prairie used in this study, thereby increasing the potential loss of topsoil and C from erosion and the loss of SOC from oxidation and decomposition. Similar to the results of this study, Brye et al. [15] reported a net loss of SOC from the top 10 cm in the Grand Prairie during a 14-year period from the same Seidenstricker sites used in this study, where, averaged by landuse, SOC decreased at a rate of 0.1 Mg SOC ha^{-1} $year^{-1}$. Over a 6-year period between 2001 and 2007, Brye and Gbur [13] also reported, averaged across landuse, soil C sequestration rates in the top 10 cm were 0.5 and −0.4 Mg SOC ha^{-1} $year^{-1}$ in the Ozark Highlands and the Grand Prairie, respectively. Similarly, Brye and Gbur [14] reported that, on average, SOC increased by 0.13 Mg SOC ha^{-1} $year^{-1}$ in the top 10 cm in the Ozark Highlands. In contrast to the results of this study, a field experiment conducted on a silt-loam soil under soybean production in the Mississippi River Delta region of eastern Arkansas, approximately 80 km east of the sites in the Grand Prairie used for this study, showed that TC in the top 10 cm increased at an average of 0.6 Mg SOC ha^{-1} $year^{-1}$ over a 6-year study period across numerous tillage-burn-residue-level treatment combinations [24]. A study conducted across Texas (e.g., Bushland, Temple, and Corpus Christi) determined the linear relationship between SOC sequestration and average annual temperature was stronger (r^2 = 0.99) than with rainfall (r^2 = 0.40), where SOC decreased by 0.17 Mg ha^{-1} $year^{-1}$ for every degree increase in the annual

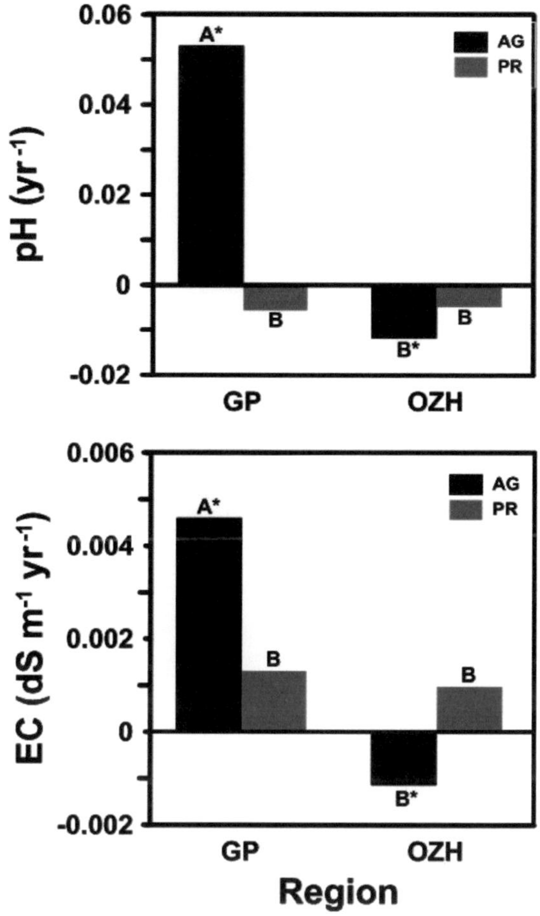

Figure 3.
Landuse effects by physiographic region on soil organic matter (SOM), total nitrogen, and total carbon content and concentration in the top 10 cm from the 2016 sampling only under either agricultural management (AG) or undisturbed prairie (PR) landuse in the Grand Prairie (GP) region of east-central Arkansas and the Ozark Highlands (OZH) region of northwest Arkansas. Different letters associated with mean values on a panel are different at P < 0.05.

temperature, and SOC only decreased by 0.0023 Mg ha^{-1} $year^{-1}$ as rainfall increased [25]. Results of the current study support the well-documented pattern that SOC sequestration is greater in cooler (i.e., the Ozark Highlands) compared to warmer climates (i.e., the Grand Prairie); however, the relationship between soil moisture levels and SOC sequestration is more difficult to predict and does not follow a linear relationship.

As a result of decreased TN and increased TC, averaged across landuse, the soil C:N ratio increased in both regions, but increased more over time in the Ozark Highlands (0.18 $year^{-1}$) than in the Grand Prairie region (0.13 $year^{-1}$, **Table 6**). Similar to the results of this study, Brye and Gbur [13] concluded that the soil C:N ratio increased between 2001 and 2007, but only under the agricultural landuse in the Grand Prairie and that the soil C:N ratio did not change over time under native prairie in the Grand Prairie and the Ozark Highlands, as well as under agricultural landuse in the Ozark Highlands. In contrast, Brye and Gbur [14] showed that the soil C:N ratio increased by 3.6% between 2002 and 2008 in grasslands in the Ozark Highlands.

Averaged across landuse, the TN fraction of SOM changed more over time in the Grand Prairie (0.25% $year^{-1}$) than in Ozark Highlands (−0.04% $year^{-1}$), but both did not differ from a change of zero (**Table 6**). In contrast, the TC fraction of

	Soil properties		
Treatment	SOM (% yr^{-1})	TN (% yr^{-1})	TC (% yr^{-1})
Region			
Grand Prairie (GP)	−0.04b*	−0.004b*	−0.03b*
Ozark Highlands (OZH)	−0.01a	−0.002a*	0.02a*
Landuse			
Agriculture (AG)	−0.04	−0.003	−0.00
Prairie (PR)	−0.01	−0.003	−0.01
Region × landuse			
GP-AG	−0.05	−0.003	−0.02
GP-PR	−0.04	−0.004	−0.04
OZH-AG	−0.03	−0.003	0.01
OZH-PR	−0.01	−0.001	0.02

**An asterisk indicates mean value is greater than 0 (P < 0.05).*

Table 7.
Summary of mean soil properties by treatment (i.e., physiographic region, landuse, and their interaction) for soil organic matter (SOM), total nitrogen (TN), total carbon (TC) concentrations in the top 10 cm from the 2016 sampling in silt-loam soils in Arkansas.

SOM, averaged across landuse, increased over time in the Ozark Highlands (0.52% $year^{-1}$), where the TC fraction of SOM decreased over time in the Grand Prairie (−0.32% $year^{-1}$) (**Table 6**). Increased TN within the available above- and below-ground biomass and the resulting SOM, especially in N-limited ecosystems such as native prairies or agricultural fields, can lead to increased microbial decomposition and therefore loss of C and N within the soil until periods of anaerobic conditions or cooler temperatures slow down decomposition process, therefore resulting in an accumulation of SOC. Brye and Gbur [14] reported that TC and TN fractions of SOM did not change over time in the Ozark Highlands, whereas the difference in results could be the result of the shorter sampling period of only 7 years used by Brye and Gbur [14] compared to the longer 15-year sampling period used in this study.

In contrast to the other measured soil properties, changes in TC content, C:N ratio, and TC fraction of SOM in the top 10 cm differed ($P < 0.05$) over time between landuses (**Table 5**). Averaged across physiographic region, the change in TC content and the TC fraction of SOM over time were greater in the agricultural compared to the native prairie landuse, but neither soil property change over time differed from a change of zero (**Table 6**). In contrast to TC content and the TC fraction of SOM, averaged across physiographic region, the C:N ratio increased more over time in the agricultural (0.19 $year^{-1}$) than in the native prairie landuse (0.12 $year^{-1}$; **Table 6**).

These results are somewhat contradictory to the stated hypothesis, where greater SOC sequestration was expected to occur in the native prairie landuse over time. However, the disagreement between the results and the stated hypothesis was likely driven by the fact that the largest numeric increase in TC content occurred within the agricultural landuse (0.32 Mg ha^{-1} $year^{-1}$) in the Ozark Highlands and the greatest numeric decrease occurred in the native prairie in the Grand Prairie (−0.47 Mg ha^{-1} $year^{-1}$). Since the agriculturally managed soils in the Ozark Highlands were not tilled on any regular basis and were used as grazed pastureland and mowed hayland, it is within reason that SOC would be increasing within these ecosystems, as there are likely fertilizers added annually or semi-annually to these agroecosystems in the form of both inorganic fertilizers and/or manure, which would stimulate plant growth and increase above- and belowground biomass to add to the SOM and SOC pools.

Managed grazing practices (i.e., rotational grazing) have also been known to increase SOC storage in soils. A grazing study conducted in a northern mixed-grass prairie in Wyoming under both light and heavy stocking rates reported increased SOC in the top 30 cm (0.30 Mg C ha^{-1} $year^{-1}$) compared to the non-grazed surrounding exclosures [26]. However, a meta-analysis conducted on C sequestration in native rangelands of the North American Great Plains revealed that, although there was no statistical relationship between the change in SOC content and the longevity of a grazing management practice, the general trend suggested a decrease in SOC sequestration as the age of the grazing management system increased, where the range of years under management were between about 20 and 80 years [27]. The duration under consistent grassland management for the sites in this study within the Ozark Highlands was roughly greater than 20 years, thus was at the younger end of the age range evaluated by Derner and Schuman [27].

The decrease in TC in the native prairie within the Grand Prairie was unexpected, where the most likely explanation was the combination of severe fragmentation and periodic vehicle traffic and compaction from agricultural machinery in order to reach the cultivated fields surrounding the prairie. Contrary to the results of this study, Brye et al. [15] reported that, in the same native prairie at the Seidenstricker site in this study, a significant increase in SOC concentration occurred from 1987 to 2001. Brye and Gbur [13] also reported that landuse differences in SOC content change over time, averaged across physiographic region, equated to SOC sequestration rates of 0.4 and −0.3 Mg ha^{-1} $year^{-1}$ in the top 10 cm of the native prairie and agricultural landuse, respectively. The loss of SOC by the conversion of natural vegetation to cultivated landuse, as well as the continued loss of SOC as duration under cultivation increases, is well known, despite varying results due to factors such as soil texture, cropping system, residue management, and climate [6, 8, 28–30]. Tillage can have one of the greatest influences on SOC loss over time due the disturbance of the macroaggregates that form around and protect particles of undecomposed SOM, leading to the mineralization of that SOM, and consequently a loss of SOC. A study conducted on a silty-clay-loam soil in south-central Texas reported an average 50% increase in SOC storage in the top 5 cm over a 20-year sampling period under a no-tillage management plan compared to a conventional tillage practice [30].

Jones and Donnelly [31] conducted a meta-analysis study among landuses ranging from native undisturbed grasslands to poorly managed rangelands and concluded global soil C sequestration rates in the top 15 and/or 30 cm ranged from 0 to approximately 8 Mg SOC ha^{-1} $year^{-1}$. In the current study, the largest mean soil C sequestration rate was measured in the agriculturally managed soils within the Ozark Highlands (0.32 Mg C ha^{-1} $year^{-1}$). However, greater SOC sequestration rates were expected to occur in the native prairies compared to the agricultural landuse and greater accumulation of SOM, TC, and TN was expected in the Grand Prairie region compared to the Ozark Highlands based on results of previous studies conducted at these sites [11–15]. Differences between studies conducted previously at these sites could stem from variations in sampling and analytical methods over time; however, differences are more likely the result of actual changes over time, identified by direct measurements, over a longer period (i.e., 15 years) in this study instead of using regression analyses or linear relationships and shorter study periods (i.e., ≤8 years) that were used in several of the previous studies [11–15].

Considering only the Grand Prairie sites consisting of a native tallgrass prairie and three agroecosystems that varied in duration under cultivation that originated as part of the native prairie tract, regression analysis revealed no significant relationship between soil C sequestration rate in the top 10 cm over the 15-year period from 2001 to 2016 ($P > 0.05$) or TC storage from 2016 alone ($P > 0.05$) and duration of years of annual cultivation. Severe fragmentation and mismanagement of the native tallgrass prairie could be the cause of this lack of a significant linear relationship. Brye et al. [15] somewhat similarly concluded that the relationship between SOC and years of cultivation did not change significantly over a 14-year period between 1987 and 2001 in the same study sites within the Seidenstricker site.

4. Conclusions

Changes in near-surface soil C and N and related properties, assessed by direct measurement, over a 15-year period in silt-loam soils in Arkansas differed between physiographic regions and landuse and among their treatment combinations. Similar to that hypothesized, averaged across region, SOM, TC, and TN in the native prairie landuse did not change over time, indicating some degree of equilibrium exists in the less-disturbed, more natural ecosystems. However, in contrast to that hypothesized, SOM, TC, and TN also did not change over time in the managed agricultural landuse when averaged across region. Though not significant, cultivated row-crop agriculture on alluvial soils was shown to have at least numerically lower C and N storage and C and N decreased more over time than that in managed pastureland on residual soils, likely due to the long history of intensive tillage, despite alluvial soils typically being generally considered more fertile than residual soils.

Results of this study demonstrate how the combination of climate and soil parent material, which constituted the major differences between physiographic regions that were investigated in this study, can have a large influence on SOM, C and N storage, and change over time. Despite differing types of managed agricultural landuse between the two regions, physiographic region clearly had a greater influence than landuse, as evidenced by more soil property changes over time evaluated in this study differing between regions when averaged across landuse than differed between landuses when averaged across regions.

Results also showed that more numerous differences between regions and landuses were identified when only a single measurement set in time was considered compared to much fewer differences between regions and landuses recognized when assessing change over time based on direct measurements. In the absence of direct measurements, any inferences drawn about temporal trends in soil properties, particularly those like SOM, C, and N, that are key to improving understanding about the effects of rising mean annual air temperatures, rising atmospheric greenhouse gas concentrations, and global climate change in general must be tempered with numerous caveats because those inferences could be misleading.

Many types of ecosystems are resilient and conditioned to resist change. Though inconvenient for numerous reasons, direct measurement over time in long-term studies, as were conducted in this study, perhaps offers the most appropriate methodology to assess temporal variation in soil properties and ecosystem characteristics toward understanding global climate change. Therefore, long-term, direct-measurement studies should be maintained and expanded to increase the accuracy of cataloging important ecosystem processes, such as soil C sequestration and other beneficial soil properties, particularly in disappearing native prairie ecosystems in Arkansas and elsewhere. The results of long-term studies will provide more useful and effective guidance for rehabilitating and/or restoring areas of degraded land and/or minimally productive agricultural land. Ecosystem restoration projects will not only likely increase soil health and sustainability, but applying similar restoration principles to agricultural lands may increase productivity and collectively contribute to slowing, or potentially reversing, the global threat of rising greenhouse gases in the atmosphere and climate change.

Acknowledgements

This research study was partially funded by the Arkansas Natural Resources Conservation Service.

Conflict of interest

There are no conflicts of interest to declare.

Author details

Marya McKee, Kristofor R. Brye* and Lisa Wood
University of Arkansas, Fayetteville, AR, USA

*Address all correspondence to: kbrye@uark.edu

References

[1] International Panel on Climate Change (IPCC). Carbon and Other Biogeochemical Cycles. Climate Change 2014—The Physical Science Basis [Online]. 2014. pp. 465-514. Available from: http://ipcc.ch/index.htm [Accessed: 25 May 2018]

[2] Environmental Protection Agency (EPA). Greenhouse gas emissions: Sources of greenhouse gas emissions [Internet]. 2017. Available from: https://www.epa.gov/ghgemissions/sources-greenhouse-gas-emissions [Accessed: 25 May 2018]

[3] Stevenson FJ. Carbon balance of the soil and role of organic matter in soil fertility. In: Stevenson FJ, editor. Cycles of Soil: Carbon, Nitrogen, Phosphorus, Sulfur, Micronutrients. New York, NY: John Wiley and Sons; 1986. pp. 45-77

[4] Brady NC, Weil RR. The Nature and Properties of Soils. 14th ed. Upper Saddle River, NJ: Pearson Education Inc; 2008. p. 980

[5] Lal R, Follett RF, editors. Soil Carbon Sequestration and the Greenhouse Effect. 2nd ed. Vol. 57. Madison, WI: Soil Science Society of America Special Publication; 2009. p. 410

[6] McCarl BA, Metting FB, Rice C. Soil carbon sequestration. Climate Change. 2007;**80**:1-3

[7] Kucharik CJ, Brye KR, Norman JM, Foley JA, Gower ST, Bundy LG. Measurements and modeling of carbon and nitrogen cycling in agroecosystems of southern Wisconsin: Potential for SOC sequestration during the next 50 years. Ecosystems. 2001;**4**:237-139

[8] Post WM, Kwon KC. Soil carbon sequestration and land-use change: Process and potential. Global Change Biology. 2000;**6**:317-327

[9] Six J, Bossuyt H, Degryze S, Denef K. A history of research on the link between (micro)aggregates, soil biota, and soil organic matter dynamics. Soil and Tillage Research. 2004;**79**:7-31

[10] United States Department of Agriculture, Soil Conservation Service (USDA-SCS). Land resource regions and major land resource areas of the United States, the Caribbean, and the Pacific Basin. In: Agriculture Handbook 296. Washington DC: USDA-SCS; 1981. p. 682

[11] Brye KR, West CP. Grassland management effects on soil surface properties in the Ozark Highlands. Soil Science. 2005;**170**:63-73

[12] Brye KR, West CP, Gbur EE. Soil quality differences under native tallgrass prairie across a climosequence in Arkansas. The American Midland Naturalist. 2004;**152**:214-230

[13] Brye KR, Gbur EE. Regional differences in soil carbon and nitrogen storage as affected by land use and soil moisture regime. Soil Science. 2010;**175**:339-348

[14] Brye KR, Gbur EE. Near-surface soil property changes over time as affected by grassland management in the Ozark Highlands. Soil Science. 2011;**176**:129-135

[15] Brye KR, Gbur EE, Miller DM. Relationships among soil carbon and physiochemical properties of a Typic Albaqualf as affected by years under cultivation. Communications in Soil Science and Plant Analysis. 2004;**34**:177-192

[16] Brye KR, Pirani AL. Native soil quality and the effects of tillage in the Grand Prairie region of eastern Arkansas. The American Midland Naturalist. 2005;**154**:28-41

[17] National Resource Conservation Service (NRCS). Web Soil Survey [Internet]. 2017. Available from: http://websoilsurvey.nrcs.usda.gov/ [Accessed: 25 May 2018]

[18] Holiman B, Woolbright J, James D. Increasing Grassland Block Size and Restoring Ozark prairie and Woodlands. Arkansas State Wildlife Grant Pre-proposal [Internet]. 2013. Available from: http://www.wildlifearkansas.com/proposals/2013PreProposals/ [Accessed: 25 May 2018]

[19] Arkansas Natural Heritage Commission (ANHC). System of Natural Areas: Natural Areas [Internet]. 2009. Available from: http://www.naturalheritage.com/NaturalAreas/natural-areas-1 [Accessed: 25 May 2018]

[20] National Centers for Environmental Information (NCEI). Data Tools: 1981-2010 Normals. 1981-2010 Normals | Data Tools | Climate Data Online (CDO) | National Climatic Data Center (NCDC) [Internet]. 2010. Available from: https://www.ncdc.noaa.gov/cdoweb/datatools/normals [Accessed: 25 May 2018]

[21] Brye KR, Riley TL, Gbur EE. Prairie restoration effects on soil properties in the Ozark Highlands. Journal of Integrated Biosciences. 2008;**4**:87-104

[22] Brye KR, Slaton NA. Carbon and nitrogen storage in a typic Albaqualf as affected by assessment method. Communications in Soil Science and Plant Analysis. 2003;**34**:1637-1655

[23] Brye KR, Riley TL. Soil and plant property differences across a chronosequence of humid-temperate tallgrass prairie restorations. Soil Science. 2009;**174**:346-357

[24] Amuri N, Brye KR, Gbur EE, Popp J, Chen P. Soil property and soybean yield trends in response to alternative wheat residue management practices in a wheat-soybean, double-crop production system in eastern Arkansas. Journal of Integrated Biosciences. 2008;**4**:64-86

[25] Potter KN, Tolbert HA, Jones OR, Matocha JE, Morrison JE, Unger PW. Distribution and amount of soil organic C in long-term management systems in Texas. Soil and Tillage Research. 1998;**47**:309-321

[26] Schuman GE, Reeder JD, Manley JT, Hart RH, Manley WA. Impact of grazing management on the carbon and nitrogen balance of a mixed grass rangeland. Ecological Applications. 1999;**9**:65-71

[27] Derner JD, Shuman GE. Carbon sequestration and rangelands: A synthesis of land management and precipitation effects. Journal of Soil and Water Conservation. 2007;**62**:77-65

[28] Lal R. Carbon sequestration impacts on global climate change and food security. Science. 2004;**304**:1623-1627

[29] Lal R. Soil carbon sequestration to mitigate climate change. Geoderma. 2004;**123**:1-22

[30] Wright AL, Hons FM. Tillage impacts on soil aggregation and carbon and nitrogen sequestration under wheat cropping sequences. Soil and Tillage Research. 2005;**84**:67-75

[31] Jones MB, Donnelly A. Carbon sequestration in temperate grassland ecosystems and the influence of management, climate and elevated CO_2. New Phytologist. 2004;**164**:423-439

Chapter 4

Strategic Management of Grazing Grassland Systems to Maintain and Increase Organic Carbon in Soils

Mohammad Ibrahim Khalil, Rosa Francaviglia, Beverley Henry, Katja Klumpp, Peter Koncz, Mireia Llorente, Beata Emoke Madari, Miriam Muñoz-Rojas and Rainer Nerger

Abstract

Understanding management-induced C sequestration potential in soils under agriculture, forestry, and other land use systems and their quantification to offset increasing greenhouse gases are of global concern. This chapter reviews management-induced changes in C storage in soils of grazing grassland systems, their impacts on ecosystem functions, and their adaptability and needs of protection across socio-economic and cultural settings. In general, improved management of grassland/pasture such as manuring/slurry application, liming and rotational grazing, and low to medium livestock units could sequester C more than under high intensity grazing conditions. Converting cultivated land to pasture, restoration of degraded land, and maximizing pasture phases in mixed-cropping, pasture with mixed-livestock, integrated forestry-pasturage of livestock (silvopastoral) and crop-forestry-pasturage of livestock (agro-silvopastoral) systems could also maintain and enhance soil organic C density (SOCρ). In areas receiving low precipitation and having high erodibility, grazing exclusion might restore degraded grasslands and increase SOCρ. Yet, optimizing C sequestration rates, sowing of more productive grass varieties, judicial inorganic and organic fertilization, rotational grazing, and other climate-resilient approaches could improve overall farm productivity and profitability and attain sustainability in livestock farming systems.

Keywords: carbon sequestration, grazing grassland, silvopastoralism, integrated land uses, livestock farming

1. Introduction

Soil stores 2–3 times more carbon (C) than the atmosphere. Soil organic carbon (SOC) pools under contrasting long-term management systems provide insights into the potential for sequestering C, sustaining soil productivity and maintaining functions in the biosphere-atmosphere interface. The broadest division of grassland, both

IntechOpen

natural and anthropogenic, is between temperate and tropical grasslands. Globally, grasslands (pasture, silage and hay) dominate major agricultural areas and contribute 20–30% to the SOC pool by sequestering atmospheric CO_2, thus mitigating climate change [1, 2]. Livestock graze mostly on pasture and meadows, and the production systems are highly diverse, ranging from low-input grasslands in arid and semiarid regions to highly intensive pasture in more mesic environments, integrating livestock-crop-forage systems. Grazing is one of the most important factors that could change the soil C density in grassland systems. Understanding the impacts of grazing intensity and livestock types under different management systems on SOC sequestration is a key to providing the most effective soil C management strategies.

Soil C storage depends on the C input mainly through the root biomass, added C, and its release mediated by soil processes. Belowground processes may respond differently from aboveground vegetation to grazing whereas change of plant community structure induced by grazing does not necessarily lead to decreased soil C storage [3]. Although grazing in some cases decreases vegetation growth, good management can improve its growth on many degraded lands [4]. In addition to biogeochemical processes and environmental factors, SOC storage under grasslands and the associated land uses is regulated importantly by biotic factors, e.g., livestock type, grazing intensity, grass species, and their heterogeneity [1]. Some recent studies show that intensification of livestock management could enhance C losses in association with emissions of GHGs [5]. Others show that intense grazing pressure and large additions of manure over short periods (e.g., rotational grazing) increase soil C and water infiltration and retention, and eventually enhance plant production [4]. Soil C sequestration in grasslands varies between 0.03 and 1 t C ha^{-1} $year^{-1}$, depending on land type, land use, climatic factors, and treatments [6].

Grazing intensity and management may modify soil physical structure, function, and SOC storage capacity that could reduce or increase nutrient retention, water storage, pollutant attenuation, soil fertility, plant productivity, and species composition [7]. For sustainable management of pastures and rehabilitation of degraded lands, tailoring flexible and site-specific grazing management, depending on climate conditions and the availability of local resources, and avoidance of the extreme process of land degradation that may deteriorate further with climate change are in need. Carbon balance is controlled by the nature, frequency, and intensity of disturbances in grassland ecosystems [8]. However, the relationship between grazing intensity and SOC is generally nonlinear [9]. Previous studies have found mixed results [10], with some showing increases [11], no effect [12], or decreases [13] in SOCρ. Other recent reviews [10] state that high grazing intensity significantly decreases belowground C and N pools, and those effects depend on livestock type and climatic conditions. However, some mechanisms are not well understood, and mixed results are common. Animal manure and other offsite organic applications have significant potential for sequestering C in soils, but the proportion stabilized may depend on local climatic and edaphic conditions and the decomposability/degradability of the materials added [14]. Grazing also accelerates N cycling and promotes N losses through NH_3 from urine and dung patches [15] which, in nutrient limited systems, may constrain C inputs and humification rates.

Globally, most soils are responsive to management changes to increase SOCρ. The greatest response comes from retirement and restoration of degraded agricultural lands, manure/bio-solid applications [10], improvements of pasture lands, adaptive grazing management systems, inclusion of woody species into the pasture system, and conversion from cropland to pasture [6]. This chapter advances evidence-based C sequestration potential in soils of pasture and associated lands under various livestock systems (**Figure 1**). The main aims are to (a) improve the knowledge base and understanding of

(a) (b) (c) (d) (e) (f)

Figure 1.
Livestock grazing in grassland and other related land uses. (a) Grazing grassland in Temperate regions (Source: ARC 2020), (b) Grazing grassland in Tropical regions (Source: SLU), (c) Grazing grassland in Mediterranean regions (Source: Dreamstime), (d) Grazing grassland in Arid regions (Source: NMSU), (e) Integrated crop-livestock system (Source: People Food & Nature) and (f) Integrated silvo-pastural system (Source: Aftaweb).

management practices and technologies to increase and maintain SOC while reducing climate change footprint and achieving productivity and environmental benefits; (b) identify region/biome-specific management practices to enhance/maintain SOC and combat environmental degradation without sacrificing food security; and (c) outline economic, ecological, social, and policy options for storing additional SOC.

2. Carbon sequestration potential in soils under livestock-associated pasture management practices

2.1 Grazing grassland/pasture management

Adjustment of grazing/management intensity to climate and soil type could increase SOC. Positive and negative impacts of grazing on SOC storage compared to unharvested rangeland have been reported for semi-arid regions of the USA [5], Western Canada [16], the Netherlands [17] and the United Kingdom [18]. In many cases, grazing favors C sequestration via animal returns and heterogeneity of vegetation with the exception of very intensive systems. This can be considered as a mosaic of patches of variable vegetation height, with or without the presence of urine and dung. In contrast, very intense grazing, or short periods between successive grazing can lead to a trade-off between biomass production and C inputs to soil (i.e., root production and litter), and subsequent C sequestration [8].

Even so, managed grasslands have the potential to act as C sinks, sequestrating on average 0.7 ± 0.16 t C ha^{-1} $year^{-1}$ [19]. However, there is a large variability in soil C accrual due to differences in climate, soil, and vegetation conditions as well as due to varying biomass removal. Under low biomass removal [~30% of present biomass], soil C sequestration of European grasslands may reach up to 1.27 ± 0.40 t C ha^{-1} $year^{-1}$, while at medium biomass removal (30–70%) to high (>70%) values are lower, depending on fertilization level and climate (**Table 1**). Indeed, high biomass removal (>70%) may lead to SOC losses. In North Dakota, USA, a long-term study in three mixed prairie sites (mainly Blue grama: *Bouteloua gracilis*) found that the moderately grazed pasture (2.6 ha $steer^{-1}$) contained 17% less SOC than the exclosure treatment, but heavy grazing (0.9 ha $steer^{-1}$) did not reduce it further (**Table 1**) [5, 18]. For 12 years, the annual rate of change in SOC (0–90 cm) followed the order: low grazing pressure (1.17 t C ha^{-1} $year^{-1}$) > unharvested (0.64) = high grazing pressure (0.51) > hayed (0.22). Moreover, grazed (cattle) tall fescue-common bermudagrass pasture (20 years old) had greater SOC (31%) at a depth of 0–20 cm than adjacent 24-year old conservation-tillage cropland [19]. Improved C sequestration for extensive grazing, showing a sink of C (0.86 ± 0.74 t C ha^{-1} $year^{-1}$), vs. mown systems was also confirmed for Hungarian sandy grasslands, during which the mowing management (cut once per year) became a source of C (−1.22 ± 0.35 t C ha^{-1} $year^{-1}$) [20]. These C losses were attributed to a higher herbage use intensity of the mowed grassland compared to the grazed one. In addition to intensive biomass removal, soil erosion can contribute to reducing SOC in pastures heavily grazed by cattle.

Even so, under some conditions, high biomass removal may improve C sequestration. Such as the semi-arid grasslands in Colorado (shortgrass steppe), where changes in SOC were higher with heavy grazing (60–75% utilization; 2.27 t C ha^{-1} $year^{-1}$) compared to light grazing (20–35% utilization; 0.55 t C ha^{-1} $year^{-1}$) (**Table 1**) [21]. Significantly higher soil C (0–30 cm) was measured in grazed pastures (1.06 ± 0.03 t C ha^{-1} $year^{-1}$) compared to nongrazed exclosures (0.20 ± 0.14). In fact, heavy stocking rates in shortgrass steppe resulted in a plant community dominated by the C4 grass, blue grama, while exclusion of livestock grazing increased the production of C3 grasses and prickly pear cactus (*Opuntia polyacantha*) [22]. In addition to plant community changes, grazing exclusion leads to an immobilization of C in excessive aboveground plant litter of forbs and grasses and lack of dense fibrous rooting systems conducive to SOM formation and accumulation. Accordingly, outcomes indicate that stimulation of annual shoot turnover and redistribution of C within the plant-soil system contributes to an increase in SOC. Furthermore, the higher SOC in heavily grazed grassland may be attributable to higher inorganic C (SIC), than the nongrazed treatment, and long-term grazing could decrease the readily mineralizable fraction of SOM. These effects emphasize the importance of inorganic C in assessing the mass and distribution of plant-soil C, and in evaluating the impacts of grazing management on C sequestration particularly in semiarid and arid ecosystems.

C sequestration, of course, should not be the only priority when making decision of pasture-based livestock farming systems located in less productive areas, (e.g., southern Europe and mountainous regions), which are highly relevant in both environmental and social terms. Although, C sequestration should be promoted to mitigate climate change and improve soil quality (water holding capacity and nutrient turnover). Solutions such as adaptive multi-paddock grazing or rotational grazing systems may increase carrying capacity and restore soil C. In other systems

Biomes/regions	Livestock category	Livestock density (LSU: L, M, and H) * and management practices	Fertilization (N, P) and liming (kg ha^{-1})	SOCρ changes (t C ha^{-1} year^{-1})
The EU and France (managed grassland) [23]	Cattle	L = LSU <0.6; HUI <0.3	Zero and ≥100 N	1.27 ± 0.40*
		M = LSU 0.6–1.3, HUI 0.3–0.7	Zero and ≥100 N	1.12 ± 0.32*
		H = HUI 0.7–1	Zero	−0.57 ± nd
			>100 kg N	0.74 ± 0.30
Australia (perennial and annual pasture) [24–28]	Mixed (cattle/ sheep)	M	—	0.50 ± nd
			Liming ± nutrients	0.40 ± 0.06*
			Liming ± phosphate	0.35 ± nd*
		Density = 0%, L to H = 50–200%	—	0.10 ± 0.10 −0.45 ± 0.53*
Rotational grazing [29]		M	—	0.35 ± nd
Hungary (extensive grazing) (Grazing +1 cut) [20]	Cattle (mowed)	L = 0.64 NLSU, HUE: 0.4	—	0.86 ± 0.74
		L = HUE 0.6	—	−1.23 ± 0.35
Hungary [30]	Mixed	L (Cattle, sheep, goat, and horses)	—	0.0013 ± nd
Mediterranean, Spain (extensive grazing)		L = 0.7–2.5 ewe eq ha^{-1} (cattle, sheep, pigs, and goats)	—	0.05–0.10
USA (mixed prairie) [5] (Grazed Bermuda grass) [31, 32] -Mixed prairie - Short-grass steppe [21, 22]	Steers	H + M	Inorg. (N: 200–270), inorg-org. (73.6), and broiler litter	0.03 ± 0.00
	Cattle Angus Steers	L H	—	1.17 ± nd 0.51 ± nd
		M	—	0.65 ± nd
	Sheep	L H	—	0.20 ± nd* 1.51 ± nd*
		L = YH (20–35% utilization) H = YH (60–75% utilization)	Long-term grazing Heavy grazing	0.55 ± nd 2.27 ± nd

*LSU = Livestock density unit (ha^{-1} year^{-1}); L = Low; M = Medium; H = High; nd = Not determined; YH = Yearling heifers; SOCρ = SOC density; * = Pooled/Averaged.*

Table 1.
Annual SOC density changes (t C ha^{-1} year^{-1}) in grazing grassland/pasture and the associated management practices.

such as marginal grasslands, introducing perennial grasses can increase pasture productivity [33] and build SOC storage, while minimizing surface erosion. Perennial grasses, compared to annuals, generally allocate a greater fraction of productivity to the maintenance of a deeper and more extensive root system [24, 25], resulting in an average increase of 0.15–0.50 t C ha^{-1} $year^{-1}$. Also, the introduction of more productive species [34], improved grazing regimes, fertilization practices, and irrigation management has been proposed for intensively managed pastures in North America for further potential C gains of 0.2 t C ha^{-1} $year^{-1}$.

Practices, including liming, gypsum amendment (e.g., 125 kg ha^{-1}), and nutrient managements, could increase SOC within a range of 0.29–0.55 t C ha^{-1} $year^{-1}$ (e.g., Australia). These gains are primarily attributed to increased plant production. In general, mineral phosphate fertilizers are applied to pastures on granite-derived soil, while gypsum is applied to address inherent deficiencies on basalt-derived soil. The application of P either alone or coupled with lime sequestrated C in soils at 0.41 and 0.29 t C ha^{-1} $year^{-1}$. However, for land occupied by low to medium intensity grazing (~90% of Australia's agricultural land), soil and climate conditions are not suitable for other more intensive agricultural practices. Given the large area occupied by these lands, a small gain in SOC per hectare would translate to a high total sequestration [26].

2.2 Integrated farming with grazing

Adoption of sustainable practices is needed to maintain soil fertility and subsequent productivity, and to avoid soil degradation and SOC depletion. Integrated farming systems often provide a combination of good management practices. The success of integrated system in promoting SOC accumulation largely depends on successful maintenance of good management practices over time. Besides, C accumulation and the capacity of the soil to maintain its levels depend on a variety of factors such as clay soils contribute to higher SOCρ and its maintenance [35, 36] and above-ground species diversity such as co-existence of shallow and deep-rooting species [37] influence below-ground diversity [38] and provide a constant soil cover and biomass inputs to combat erosion and maintain nutrient inputs balance at various soil depths through.

Incorporating legumes into grazed grasslands and woodlands/savannas can address the N deficiency as common in mature, unfertilized rangeland soils. This practice has been used in tropical/subtropical regions of northeastern Australia [39] and in areas of Western Australia with a temperate climate. Here, growing *Leucaena leucocephala* in rows in C4 grassland in subtropical regions increased SOC by 17–30% over 40 years (sequestration rate of 0.28 t ha^{-1} $year^{-1}$, **Table 2**). Also, improved management practices (fertilization, liming, irrigation, seeding legumes, planting more productive grasses, and using appropriate grazing regimes) have been reported to increase C sequestration (0.72 t C ha^{-1} $year^{-1}$) in 22 municipalities of the Brazilian states of Rondônia and Mato Grosso [40].

A number of integrated farming systems (IS) such as crop-livestock (agropastoral system, ICL), crop-forestry (silvoarable system, ICF), forestry-pasturage of livestock (silvopastoral system, ILF), and crop-forestry-pasturage of livestock (agro-silvopastoral system, ICLF) are reported to improve C sequestration. For IS in Southern Amazon and Cerrado (neo-tropical savanna) of Brazil, C sequestration rates of 0.60 and 1.30 t C ha^{-1} $year^{-1}$ were reported for 0–30 and 0–100 cm soil depth, respectively (**Table 2**) [37]. In the Mediterranean area of Italy with Dystric Cambisols, the conversion of cork oak forests to grasslands (i.e., silvopastoral ecosystems) showed that the C sequestration rate in topsoil (20 cm depth) was 0.71 ± 0.13 under frequent crop rotation (with 5 years of cereals or legumes, oats,

Italian ryegrass, and annual clovers or vetch followed by spontaneous herbaceous vegetation in the sixth year), and 1.20 ± 0.07 t C ha^{-1} $year^{-1}$, under temporary pasture (5 years of spontaneous herbaceous vegetation and 1 year of hay crop), respectively (**Table 2**) [41].

In agrosilvopastoral system of the Iberian Peninsula (Mediterranean woodlands, Dehesas or Montados in Spain and Portugal, respectively), the improvement of C sequestration was mainly attributed to permanent pastures with mixed livestock raising at low stocking densities without external fodder inputs, exploitation of holm and/or cork oaks and arable systems with long rotations, and closed nutrient cycles. This was in opposite to soil tillage required in many Mediterranean soil/climate conditions to allow production of arable crops likely to reduce C sequestration.

Biomes/regions	**Livestock category**	**Livestock density and/or No. LSU***	**Fertilization (N and P) and liming (kg ha^{-1})**	**SOCρ changes (t C ha^{-1} $year^{-1}$)**
Queensland, Australia (grassland and planted tree legumes mixed) [39]	Cattle	L to M = 0.45 AE ha^{-1}; 1 AE = 400 kg steer	P = 22 S = 28	0.28 ± 0.00
Southern Amazon, Brazil (integrated crop-livestock-forestry) [37]	Cattle	H = 21.27 AU ha^{-1}	370 NPK; 318 SP; 105 KCl; 324; NPK; 86 urea +10 KCl; 400 SP + 69 KCl; Pasture: 30 SP	0.60 ± 0.12 1.30 ± 0.23
Amazon, Brazil (nominal)	Beef cattle	M	—	−0.03 to 0.72 ± nd
Improved with legume and productive varieties [40]			Fertilization, lime, and irrigation	0.61 ± nd
Mediterranean, Italy (silvopasture) [41]	Sheep	L = 3–4 sheep ha^{-1}	50–39 (N-P)	0.71 ± 0.13
		L = 6 sheep ha^{-1}	50–39 (N-P)	1.20 ± 0.07
Inner Mongolia (semi-arid steppe) [9]	Cattle/sheep grazing exclusion	L	Season long grazing	0.10 ± 0.00
China (degraded grassland) [42]		M	0–30 cm 0–100 cm	0.23 ± 0.03 0.19 ± 0.04
Northern China (semi-arid, grassland) [43]		M	—	0.10 ± 0.00
China (desert steppe) [44]		H	—	1.43 ± 0.00

LSU = Livestock density unit (ha^{-1} $year^{-1}$); L = Low; M = Medium and H = High; SOCρ = SOC density change; nd: Not determined.

Table 2.
Annual SOC density changes (t C ha^{-1} $year^{-1}$) in integrated farming with grazing grassland, shrublands, and the associated management practices.

2.3 Grazing shrublands and exclusions

Overgrazing is one of the main causes of desertification in semiarid grasslands, and grazing exclusion (GE) is an effective management practice globally, to restore degraded grasslands and improve SOCρ significantly via plant biomass and soil microbial biomass compared to grazing management [44]. The C dynamics in grassland ecosystems with GE showed a positive impact of GE on vegetation and SOCρ at most sites [42]. The mean values for SOCρ change were 0.23 ± 0.03 and 0.19 ± 0.04 t C ha^{-1} $year^{-1}$ in 0–30 and 0–100 cm, respectively. Changes in SOCρ rates showed an exponential decay trend since GE and reached steady state at a later stage. Also, reduction in grazing pressure was reported (medium grazing, cattle/sheep) to result in a considerable increase of SOC and that the rate due to GE (10 years) was 0.10 t C ha^{-1} $year^{-1}$ (**Table 2**), suggesting that degradation of the grassland is being reversed [43].

For instance, the Alexa desert steppe has been strongly degraded by overgrazing, contributing around 22% of the total springtime dust originating from Asia. The effects of 7 years of GE on C dynamics showed lower SOC and higher SIC pools in areas with GE [44]. The total C pool in the GE plant-soil system was 10% greater than that in the area grazed over that time period (primarily due to 21% greater SIC), with a sequestration of 1.43 t C ha^{-1} $year^{-1}$ (**Table 2**). In semiarid steppes, recovery of heavily declined SOC caused by overgrazing is difficult and influences of long-term grazing on depression of nutrient cycling could be observed. For instance, in the semiarid steppes typical of Inner Mongolia, SOCρ was found comparable between grazed sites (average of 3 locations with sheep) and nongrazed GE sites at 0.1 t C ha^{-1} $year^{-1}$ (**Table 2**) [9]. Soil organic C levels in *Artemisia frigida* grassland was about 70% of that in *Leymus chinensis*, and in *A. frigida* grassland, it was significantly lower in grazing compared to nongrazing sites.

However, contrasting effects of overgrazing have also been reported [43, 45]. Soil IC pools could markedly contribute to the total C pool following GE, possibly due to the enhanced formation of pedogenic carbonates, higher soil water content, or increased carbonate capture in dust by recovering vegetation. On the other hand, the pool of SOC can be decreased by 11.5% in exclosure soils compared with the grazed site [46], linked mainly to the decrease in surface soil bulk density. These findings are potentially important because the Inner Mongolia grassland is the largest in the world and its degradation under heavy grazing is a source of dust storms that have major regional and global impacts. The positive impact of GE on vegetation and SOCρ [42] could improve enzyme activities and basal soil respiration in degraded sandy grassland, suggesting that degradation of the grassland is being reversed [13, 43]. A viable option for sandy grassland management could be to adopt proper exclosure in a rotation grazing system in the initial stage of degradation.

In Mediterranean regions, livestock take advantage of shrubs and grass during grazing while producing and dispersing manure. This represents the acceleration of transformation of plant OM into SOM, leading to an increase in SOCρ and makes them more resilient to future scenarios of global change [14]. Revegetation is very important especially to stop desertification processes and thereby increase SOC storage and improve soil health. Grazing management and cultivation of fodder have a high level of structural diversity both on a within and between habitat scale [47]. Greater SOCρ directly underneath the tree canopy suggests that maintaining or increasing tree cover may increase long term storage of soil C in Mediterranean silvopastoral systems.

2.4 Land use change from pasture to cropland and vice-versa

Multiple factors such as soil properties, land management, vegetation types, LUC, and climatic conditions influence soil C sequestration. Overall, the conversion of pasture into arable land leads to SOC loses with a balance up to 50 Pg [6] and up to 59% [48]. In the humid temperate zone of Europe, conversion of permanent grassland/pasture to arable land by plowing is associated with high SOC losses, especially in the first year after conversion (**Table 4**) [49].

Carbon loss after grassland conversion to cropland are often rapid (−36 ± 5%; on average estimated 1.81 t ha^{-1} $year^{-1}$) with a new SOC equilibrium being reached after 17 years [50]. Conversely, grassland establishment on cropland can be a long-lasting C sink with a relative density change of 128 ± 23% with no new equilibrium reached within 120 years (**Table 3**). Comparing sites, SOC increased with temperature and precipitation but decreased with depth and clay content. Regarding depth, top and subsoil SOC changes follow the same trend, but changes are often smaller in subsoil (25 ± 5% of the total SOC changes). Results of a meta-analysis from 74 publications indicate that overall, SOCρ declines after land use change from pasture to crop (−59%; −19 ± 7 t ha^{-1}) and increases when converting crop to pasture (+19%; 18 ± 11 t ha^{-1}; on average 0.56 t ha^{-1} of 32 years; **Table 4**) [48, 50], while mean SOC changes mostly occurred in the upper 30 cm.

Biomes/regions	Livestock category	Livestock density (LSU*)	Fertilization (N and P); liming (kg ha^{-1}) and slurry (t C ha^{-1} $year^{-1}$)	SOCρ changes (t C ha^{-1} $year^{-1}$)
Germany (grassland to cropland) [49]	Cattle	M = 2.2, 0–1 cut	~46 N, ~7 P (7 yrs); Slurry: ~0.1–0.2 t C ha^{-1} $year^{-1}$	−2.77 ± 1.79[a]
		M = 1.9; 2–3 cuts	~64 N (1 year); Slurry: ~0.2–0.4	−27.2 ± 11.70[b]
Ireland [51]	—	—	87–316 N (mineral and organic fertilizers)	−12.88[c]
Temperate zone [50]	—	—	—	−1.81 ± 0.55[d]
Europe [52]	—	—	—	−19.00 ± 7.00
Europe (cropland to pasture) [50–53]	—	—	—	1.99 ± 0.55[e]
				0.56 ± 0.34[f]
				0.22 ± nd
Australia (cropland to pasture) [2, 37, 54]	—	—	—	0.33–0.70 ± nd

LSU = Livestock density unit (ha^{-1} $year^{-1}$); L = Low; M = Medium and H = High; SOCρ = SOC density change; nd: Not determined.
[a]7-year average.
[b]1-year average.
[c]2.5-year average.
[d]20-year average.
[e]20-year average but reaching an equilibrium may take >100 years.
[f]32-year average, yet to reach equilibrium.

Table 3.
Annual SOC density changes (t C ha^{-1} $year^{-1}$) in land uses change from grazing grassland/pasture to cropland, from cropland to grazing grassland/pasture, and the associated management practices.

2.5 Degraded and other grassland areas

Soil carbon can be lost from grassland areas due to degradation, sometimes in association with management for grazing. For example, evaluation of the effects of management on SOCρ in grasslands compared to native vegetation in 22 municipalities of the Brazilian states of Rondônia and Mato Grosso showed that in degraded grassland, SOCρ declined by about 0.27–0.28 t C ha^{-1} $year^{-1}$ [39]. This indicates that degraded unmanaged grasslands in tropical regions lose C from the system (**Table 4**). Similar losses were found in temperate regions. In a shortgrass steppe, 28% less SOC was measured at locations with little or no plant input for about 45 years [55].

To assess the impact of animal trampling on soil properties, a study was conducted at a sub-alpine pasture in the Canton of Fribourg, Switzerland that had been used for summer grazing for over 150 years with a livestock density of about 4 cattle ha^{-1} (**Table 4**) [56]. The SOCρ in the 0–25 cm depth for areas of intensive trampling ("Bare steps") was 60 t C ha^{-1}, vegetated shoulder between trampled areas 74 t C ha^{-1}, and unaffected slope 76 t C ha^{-1}. The loss of SOC by trampling accounted for about 15 t C ha^{-1}, 20% of the total stock in this layer, or 30% on an equivalent soil mass basis (16 t C ha^{-1} over 150 years = 0.11 t C ha^{-1} $year^{-1}$). In the bare steps, physical protection and aggregate stability are reduced, exposing soils to the eroding power of raindrops, making them susceptible to overland waterflow, and depletion of organic C and N. Decrease in SOCρ is most probably the result of three different processes, (i) erosion of the unprotected bare soils, (ii) reduced C input due to the lack of vegetation, and (iii) soil aggregate disruption through trampling.

Sequestration estimates for more marginal and less-managed rangelands generally fall below 0.5 t C ha^{-1} $year^{-1}$ [7]. With recommended management, it was assessed that rangelands could sequester SOC at a rate of 0.1 t C ha^{-1} $year^{-1}$ with an additional 0.2–0.3 t C ha^{-1} $year^{-1}$ mitigation through avoided emissions [57]. In addition to a general increase in sequestration rates, planting of permanent vegetation in degraded and marginal lands could act as larger C sinks with sequestration rates in soils from nil to 1.1 t C ha^{-1} $year^{-1}$ depending on the use of manure/bio-solid applications [7, 15]. In summary, sequestration potential for numerous management practices could be 0.44 ± 0.20 t C ha^{-1} $year^{-1}$ [15]. While most studies have shown increased sequestration rates with improved grassland management, nil or negative effects on C sequestration in soils have been reported, possibly associated with poor experimental design or climate and soil limitations [2].

Biomes/regions	Livestock category	Livestock density (LSU*) and management practices	SOCρ changes (t C ha^{-1} $year^{-1}$)
Switzerland (summer grazing pasture) [56]	Cattle	L = Bare steps = 4 cattle ha^{-1} L = Vegetated shoulder = 4 cattle ha^{-1} L = Unaffected slope = 4 cattle ha^{-1}	−0.10 ± nd 0.09 ± nd 0.11 ± nd
Brazil (Amazon) [40]	Beef cattle	—	−0.28 ± nd

LSU = Livestock density unit (ha^{-1} $year^{-1}$); L = Low; nd: Not determined.

Table 4.
Examples of SOC density changes (t C ha^{-1} $year^{-1}$) in degraded and other land areas and the associated management practices.

3. Challenges and opportunities to improve SOC storage in pasture management systems

3.1 Livestock management impacts on soil carbon sequestration potential

Anthropogenic land use decreases soil C storage worldwide and often contributes to soil degradation and erosion. Loss of SOC occurs due to reduced organic matter inputs, deforestation, plowing, and sealing and animal impacts such as trampling. Globally, grazing lands comprise the largest and most diverse single land resource and represent an important component of terrestrial C cycling and sequestration [1]. Trade-offs among productivity, GHG emissions and SOC sequestration should be considered for the management of livestock farming to ensure sustainable production and climate mitigation. Climatic variables, particularly rainfall and temperature, and soil characteristics are major factors in determining potential storage of C in soils under managed livestock systems [58]. The data available suggest that C content varies widely among different grassland types as well as livestock management practices.

The effects of grazing management on the ecosystem processes that control C cycling and distribution have not been sufficiently evaluated in native grassland ecosystems. Current literature suggests no clear general relationships between grazing management and C sequestration. Some studies have reported no effect of grazing on SOC [43] while several others reported increases [13, 59] and a few reported decreases [42] as a result of grazing. Land use change is a major factor in determining SOC density in agricultural systems [52]. Conversion of permanent grassland or pasture into cropland results in loss of SOC, with the rate of decline dependent mainly on soil type, climate, ecosystem productivity, plant species, and intensity of management [60]. In addition, ecosystem function is affected through altered biodiversity and soil quality, with impacts differing across biomes and continents. In the tropics, the extent of degradation is normally greater due to higher temperature and often less sustainable soil management practices that accelerate decomposition and nutrient loss [60]. Similarly, conversion of native pasture and forest soils into cropland may increase soil bulk density (16%), plasticity index (30%), and soil erodibility (51%), as well as decrease SOC (50%), total N (50%), tilth index (40%), and available water capacity (40%) for surface soils [61]. There is a potential for restoration to a higher SOC level over time if arable lands are reverted to pasture. In the Mediterranean region, the recent dramatic changes in the development of industrial and tourism economies, with alteration of the composition and spatial structure of the traditional landscape, have had critical consequences for soil processes and management. European agricultural policies and the growing population have promoted intensive production systems, showing a negative effect on SOC storage in areas like Southern Spain.

3.2 Cultural and socio-economic views of grassland management

Livestock farming systems differ widely in terms of their use of resources, degree of intensification, species and orientation of production, local/regional socio-economic and market context, cultural roles, etc. [15]. Pasture-based livestock farming systems in the European Mediterranean basin play a key role in the management and conservation of large high nature value lands and that are highly relevant in both environmental and social terms, with great ecological, landscape, and cultural diversity. Accordingly, agricultural policies have begun to recognize their productive, environmental, and societal functions [62]. In the second half of the twentieth century, modernization and intensification of agriculture and the establishment of new economic and commercial relationships with urban areas have caused depopulation and a continuous reduction or abandonment of livestock farming in rural areas across Europe [63]. Within this context, the continuity of small family farms is a key aspect when assessing the sustainability of agropastoral systems. Within the EU, approximately 74 M ha of permanent grassland (including 17 M ha in upland areas), 10 M ha of temporary grassland, and 35 M ha of land in forage cereal crops (equal to 60% of the total planted area) are dedicated to feeding the European livestock herd [64]. Though grass-based systems require more land area than poultry or swine, ruminants can make use of grasslands and rangelands or land unsuitable for cultivation, thus not competing with biomass production for human food.

There are potential risks and benefits of diverse grazing land management due to the numerous episodes of land degradation associated with drought and overgrazing [27]. There is strong need for adopting sustainable practices at lower intensification management to prevent and avoid further soil degradation [65]. Implementation of land management practices with positive C storage outcomes may have a large impact on the economic factors of livestock production and can be limited by social and cultural issues. It has been suggested that reducing stocking rates to improve perennial grass basal cover could sequester 315 M t of C in the top 10 cm of soil over a 30-year period [66]. Besides, allocation of more time to grazing in pastures, rather than to feeding on mown herbage could be beneficial. Possibilities to expand grazing areas and period should also be explored. Abandonment of grasslands should be avoided to enrich biodiversity and limit the spread of invasive species [67].

Moreover, many farms are technically producing organic meat, although not yet officially accredited due to the high administrative procedures required [68]. Expansion of grazing management is possible but its feasibility and the full climate change mitigation potential in time and space on broader ecological, socio-economic, and political aspects should be evaluated [20]. The achievement of integrated systems promoting SOC accumulation and maintenance largely depends on good management practices successfully followed over time. Grasslands are among the ecosystems with the highest SOC density and stocks, and that serious concern is imperative [69]. Indeed, the EC Regulation1782/03 has introduced the concept of cross-compliance, with direct payments for farmers if they meet specific environmental requirements (Good Agricultural Environmental Conditions—GAEC).

3.3 Possible synergies and co-benefits or conflicts of livestock management with other practices

The emerging environmental and resource vulnerabilities may help adjustments in land use and farm practices, motivation and application of cultural beliefs, and a broader understanding of economic value associated with markets, technology, and administrative and policy frameworks. Importantly, we should consider the SOC stock in permanent grassland and the losses due to land use change and allocate C emission to milk and to the other products and ecosystem services to reduce the emission of GHGs from extensive systems [15]. Extensive livestock farming and pastoralism is a synergistic practice with the FAO recommendations to reduce meat consumption per capita, that is, to eat less meat but of higher quality and in the context of environmental sustainability [15]. On average, carbon emission is 4 times higher and erosion prevention is 10% lower in areas with a high grazing intensity compared to areas with a low grazing intensity [70].

Site-specific management practices could play a key role to moderate intensification of grassland production systems [71], and to mitigate some environmental impacts resulting from intensive agriculture. There are enormous opportunities across landscape mosaics to achieve an equilibrium between crops and grasslands to optimize trade-offs between food production and environment preservation as well as for offsetting unavoidable enteric CH_4 by C sequestration in grassland soils [72]. Integration of livestock systems with other agricultural activities (e.g., production of grain crops or biomass for energy) increases diversity within agricultural systems, better regulates biogeochemical cycles, decreases environmental fluxes, and supports diverse habitats and trophic networks. Thus, modern integrated crop-livestock-forestry systems could improve many ecosystem services, such as C sequestration, and environmental sustainability.

Trampling by livestock may reduce the cover and connectivity of plant, litter, and biocrust cover [73] and make soil vulnerable to wind and water erosion [74]. Strategic integration of crops, livestock, forage, and agroforestry increases the complexity of production systems, thus reducing problems of specialization, captures economic and ecological synergies, and offers a range of novel opportunities to conduit natural processes, from carbon sequestration and site remediation to nonchemical vegetation management. Mixed cropping-pasture rotation systems are likely to be significant in increasing SOC at low N application during cropping phases.

While increased complexity likely brings about ecological and economic benefits over highly specialized production systems, it may hamper its adoption and long-term maintenance. Land tenure, common property, and privatization issues also control competition from cropping, including biofuels and other land uses that limit grazing patterns and areas. Technical support for producers is imperative for the continued practice of mixed production, particularly for small- and medium-scale farmers, as well as sound complementary policies and good governance so that a "rebound effect" does not lead to any social and environmental impacts. Public extension services, in collaboration with the private sector, that strengthen information flow and enable investment in infrastructure have been and remain crucial to the success of integrated systems [37].

4. Conclusions

Globally, grazing is the largest anthropogenic land use, and a clear understanding of potential impacts of livestock management practices is essential to sequester C in the soils. Over-grazing decreases productivity, feeding efficiency, and C sequestration and increases GHG emissions from the systems. To enhance SOCρ and attain environmental sustainability in the systems, improved livestock management and the associated measures to cover soils, maintain biodiversity, select appropriate grazing time, control animal density and trampling, distribute dung and urine properly, keep soil moisture favorable, and improve livestock quality and productivity could have huge benefits. Further measures to make the system a C sink to offset any increased GHGs are to (i) optimize stocking rates to reduce land degradation, (ii) introduce improved pasture species and legumes to increase biomass production and SOCρ, (iii) apply recommended rate of inorganic fertilizers and manure to stimulate biomass production, and (iv) bring degraded land under pasture to reduce erodibility.

It is vital to understand the spatial pattern of livestock grazing intensity and its effects on ecosystem functions, address the range of natural resources and social dimensions, and encourage holistic approaches and partnership processes to achieve effective sustainable livestock-based systems. These include adaptation to climate change, promotion of technically advanced management options, introduction of consistent policies to enhance development capacity, and minimization of desertification, drought, and loss of biodiversity. Besides, recognition, awareness, behavioral change, and investments with due worldwide pro-extensive livestock policies are essential. Involvement of stakeholders, various organizations, development agencies/practitioners, community donors, relevant networks, and researchers is desirable to fully exploit the opportunities lying within the systems. Emphasis on spatially explicit global studies should be given to explore the impacts of livestock managements, adopt better technologies, and quantify trade-offs/off-setting potential and synergies among ecosystems. Further research should be targeted to value natural grasslands and livestock-based ecosystems, developing quantifiable methods for SOC measurement, strategic monitoring and verification of management-induced C sequestration. This is to ensure full GHG accounting and balance while generating improved understanding of the economic and institutional aspects of C sequestration involving people engaged in livestock farming.

Acknowledgements

The authors thank the Environmental Protection Agency (EPA), Ireland, for funding (2015 CCRP-FS.21) to publish this chapter; Department of Agriculture, Food and the Marine (DAFM), Ireland for funding the lead author through Irish Land UsES (15/S/650); and Dr. Ciniro Costa-Junior, Institute for Agriculture and Forestry Certification and Management, Piracicaba-SP, Brazil for his initial contribution.

Author details

Mohammad Ibrahim Khalil[1,2]*, Rosa Francaviglia[3], Beverley Henry[4], Katja Klumpp[5], Peter Koncz[6,7], Mireia Llorente[8], Beata Emoke Madari[9], Miriam Muñoz-Rojas[10,11] and Rainer Nerger[12]

1 School of Biology and Environmental Science, University College Dublin, Ireland

2 Climate-Resilient Agri-Environmetal Systems (CRAES)-Earth Institute, University College Dublin, Ireland

3 Council for Agricultural Research and Economics, Research Centre for Agriculture and Environment (CREA-AA), Rome, Italy

4 Science and Engineering Faculty, Queensland University of Technology, Brisbane, Australia

5 Grassland Ecosystem Research Unit, INRA, Clermont-Ferrand, France

6 Duna-Ipoly National Park Directorate, Budapest, Hungary

7 MTA–SZIE Plant Ecology Research Group, Gödöllő, Hungary

8 Forest Department, University of Extremadura, Plasencia Campus, Spain

9 Embrapa Rice and Beans, Santo Antônio de Goiás, GO, Brazil

10 School of Biological, Earth and Environmental Sciences, The University of New South Wales, Sydney, Australia

11 School of Biological Sciences, The University of Western Australia, Crawley, Australia

12 Soil and More Impacts B.V., German Office, Hamburg, Germany

*Address all correspondence to: ibrahim.khalil@ucd.ie

References

[1] Lal R. Soil carbon sequestration impacts on global climate change and food security. Science. 2004;**304**:1623-1627

[2] Conant RT, Paustian K, Elliot ET. Grassland management and conversion into grassland: Effects on soil carbon. Ecological Applications. 2001;**11**:343-355

[3] Allard V et al. The role of grazing management for the net biome productivity and greenhouse gas budget (CO_2, N_2O, and CH_4) of semi-natural grassland. Agriculture, Ecosystems and Environment. 2007;**121**:47-58

[4] Roser M, Ritchie H. Yields and Land Use in Agriculture [Internet]. 2018. Available from: https://ourworldindata.org/yields-and-land-use-in-agriculture [Accessed: 01 May 2018]

[5] Frank AB, Tanaka DL, Hoffman L, Follet RF. Soil carbon and nitrogen of northern Great Plains grasslands as influenced by longterm grazing. Journal of Range Management. 1995;**48**:470-474

[6] Janzen HH. The soil carbon dilemma: Shall we hoard it or use it? Soil Biology and Biochemistry. 2006;**38**:419-424

[7] Smith P et al. Greenhouse gas mitigation in agriculture. Philosophical Transactions of the Royal Society B: Biological Sciences. 2008;**363**:789-813

[8] Soussana J-F, Lemaire G. Coupling carbon and nitrogen cycles for environmentally sustainable intensification of grasslands and crop–livestock systems. Agriculture, Ecosystems and Environment. 2014;**190**:9-17

[9] Cui XY et al. Effect of long-term grazing on soil organic carbon content in semiarid steppes in Inner Mongolia. Ecological Research. 2005;**20**:519-527

[10] Savory A. A Global Strategy for Addressing Climate Change. 2008. Available from: https://soilcarboncoalition.org/files/globalstrategy.pdf [Accessed: 01 May 2018]

[11] Soussana J-F, Fuhrer J, Jones M, Van Amstel A. The greenhouse gas balance of grasslands in Europe. Agriculture, Ecosystems and Environment. 2007;**121**:1-4

[12] Eldridge DJ, Delgado-Baquerizo M. Continental-scale impacts of livestock grazing on ecosystem supporting and regulating services. Land Degradation and Development. 2017;**28**:1473-1481

[13] Golluscio RA et al. Sheep grazing decreases organic carbon and nitrogen pools in the Patagonian Steppe: Combination of direct and indirect effects. Ecosystems. 2009;**12**:686-697

[14] Piñeiro G et al. Pathways of grazing effects on soil organic carbon and nitrogen. Rangeland Ecology & Management. 2010;**63**:109-119

[15] Hutchinson JJ, Campbell CA, Desjardins RL. Some perspectives on carbon sequestration in agriculture. Agricultural and Forest Meteorology. 2007;**142**:288-302

[16] Naeth MA et al. Grazing impacts on litter and soil organic matter in mixed prairie and fescue grassland ecosystems in Alberta. Journal of Range Management. 1991;**44**:7-12

[17] Hassink J. Effects of soil texture and grassland management on soil organic C and N and rates of C and N mineralization. Soil Biology and Biochemistry. 1994;**26**:1221-1231

[18] Bardgett RD, Leemans DK, Cook R, Hobbs PJ. Seasonality of the soil biota

of grazed and ungrazed hill grasslands. Soil Biology and Biochemistry. 1997;**29**:1285-1294

[19] Franzluebbers AJ, Stuedemann JA, Schomberg HH, Wilkinson SR. Soil organic C and N pools under long-term pasture management in the Southern Piedmont, USA. Soil Biology and Biochemistry. 2000;**32**:469-478

[20] Koncz P et al. Extensive grazing in contrast to mowing is climate friendly based on the farm-scale greenhouse gas balance. Agriculture, Ecosystems and Environment. 2017;**240**:121-134

[21] Reeder JD, Schuman GE. Influence of livestock grazing on C sequestration in semi-arid mixed-grass and short-grass rangelands. Environmental Pollution. 2002;**116**:457-463

[22] Reeder JD, Schuman GE, Morgan JA, Lecain DR. Response of organic and inorganic carbon and nitrogen to long-term grazing of the shortgrass steppe. Environmental Management. 2004;**33**:485-495

[23] EFDC (European Fluxes Database Cluster). European Fluxes Database Cluster [Internet]. 2018. Available from: http://www.europe-fluxdata.eu/ [Accessed: 15 April 2018]

[24] Sanderman J, Farquharson R, Baldock J. Soil Carbon Sequestration Potential: A Review for Australian Agriculture. A Report Prepared for Department of Climate Change and Energy Efficiency. Canberra: CSIRO; 2010. p. 81

[25] Chan KY et al. Soil carbon stocks under different pastures and pasture management in the higher rainfall areas of south-eastern Australia. Australian Journal of Soil Research. 2010;**48**:7-15

[26] Chan KY et al. Soil carbon dynamics under different cropping and pasture management in temperate Australia: Results of three long-term experiments. Soil Research. 2011;**49**:320-328

[27] Hill MJ et al. Analysis of soil carbon outcomes from interaction between climate and grazing pressure in Australian rangelands using Range-ASSESS. Environmental Modelling & Software. 2006;**21**:779-801

[28] Segoli M et al. Managing cattle grazing intensity: Effects on soil organic matter and soil nitrogen. Soil Research. 2015;**53**:677-682

[29] Fulkerson WJ, Slack K, Moore K, Rolfe C. Management of *Lolium-perenne Trifolium-repens* pastures in the subtropics. 1. Effect of defoliation interval, seeding rate and application of N and lime. Australian Journal of Agricultural Research. 1993;**44**:1947-1958

[30] Kis-Kovács G et al. National Inventory Report for 1985-2015. Budapest: Hungarian Meteorological Service; 2017

[31] Franzluebbers AJ, Stuedemann JA. Soil-profile organic carbon and total nitrogen during 12 years of pasture management in the Southern Piedmont USA. Agriculture, Ecosystems and Environment. 2009;**129**:28-36

[32] Franzluebbers AJ, Stuedemann JA, Schomberg HH, Wilkinson SR. Soil organic C and N pools under long-term pasture management in the Southern Piedmont, USA. Soil Biology and Biochemistry. 2000;**32**:469-478

[33] Witham N et al. Evaluating Perennial Pastures: A Case Study of Perennial Pasture Use in the South Coast Region of Western Australia. Esperance, W.A.: Esperance Regional Forum; 2007. p. 86

[34] Bruce JP, Frome M, Haites E, Janzen H, Lal R, Paustian K. Carbon sequestration in soils. Journal of Soil and Water Conservation. 1999;**54**:382-389

[35] Zinn YL, Lal R, Bigham JM, Resck DVS. Edaphic controls on soil organic carbon retention in the Brazilian Cerrado: Texture and mineralogy. Soil Science Society of America Journal. 2007;**71**:1204

[36] Cardinael R et al. Increased soil organic carbon stocks under agroforestry: A survey of six different sites in France. Agriculture, Ecosystems and Environment. 2017;**236**:243-255

[37] Oliveira JM et al. Integrated farming systems for improving soil carbon balance in the southern Amazon of Brazil. Regional Environmental Change. 2018;**18**:105-116

[38] Lisboa FJG et al. The match between microbial community structure and soil properties is modulated by land use types and sample origin within an integrated agroecosystem. Soil Biology and Biochemistry. 2014;**78**:97-108

[39] Conrad KA et al. The sequestration and turnover of soil organic carbon in subtropical leucaena-grass pastures. Agriculture, Ecosystems and Environment. 2017;**248**:38-47

[40] Maia SMF, Ogle SM, Cerri CEP, Cerri CC. Soil organic carbon stock change due to land use activity along the agricultural frontier of the Southwestern Amazon, Brazil, between 1970 and 2002. Global Change Biology. 2009;**16**:2775-2788

[41] Francaviglia R, Renzi G, Ledda L, Benedetti A. Organic carbon pools and soil biological fertility are affected by land use intensity in Mediterranean ecosystems of Sardinia, Italy. Science of the Total Environment. 2017;**599-600**:789-796

[42] Deng L, Zhou-Ping S, Wu G-L, Xiao-Feng C. Effects of grazing exclusion on carbon sequestration in China's grassland. Earth-Science Reviews. 2017;**173**:84-95

[43] Su YZ, Li YL, Cui JY, Zhao WZ. Influences of continuous grazing and livestock exclusion on soil properties in a degraded sandy grassland, Inner Mongolia, Northern China. Catena. 2005;**59**:267-278

[44] Niu D et al. Grazing exclusion alters ecosystem carbon pools in Alxa desert steppe. New Zealand Journal of Agricultural Research. 2011;**54**:127-142

[45] Steffens M, Kölb A, Totsche KU, Kögel-Knabner I. Grazing effects on soil chemical and physical properties in a semiarid steppe of Inner Mongolia (P.R. China). Geoderma. 2008;**143**:63-72

[46] Berg WA, Bradford JA, Sims PL. Long-term soil nitrogen and vegetation change on Sandhill rangeland. Journal of Range Management. 1997;**50**:482-486

[47] Aguilera E, Lassaletta L, Gattinger A, Gimeno BS. Managing soil carbon for climate change mitigation and adaptation in Mediterranean cropping systems: A meta-analysis. Agriculture, Ecosystems and Environment. 2013;**168**:25-36

[48] Guo LB, Gifford RM. Soil carbon stocks and land use change: A meta-analysis. Global Change Biology. 2002;**8**:345-360

[49] Nerger R, Beylich A, Fohrer N. Long-term monitoring of soil quality changes in Northern Germany. Geoderma Regional. 2016;**7**:239-249

[50] Poeplau C et al. Temporal dynamics of soil organic carbon after land-use change in the temperate zone–carbon response functions as a model approach. Global Change Biology. 2011;**17**:2415-2427

[51] Necpalova M, Li D, Lanigan G, Casey IA, Burchill W, Humphreys J. Changes in soil organic carbon in a clay loam soil following ploughing

and reseeding of permanent grassland under temperate moist climatic conditions. Grass and Forage Science. 2014;**69**:611-624

[52] Poeplau C, Don A. Sensitivity of soil organic carbon stocks and fractions to different land-use changes across Europe. Geoderma. 2013;**192**:189-201

[53] Jones MB. Potential for carbon sequestration in temperate grassland soils. In: Abberton M et al, editors. Proceedings of the Workshop on the Role of Grassland Carbon Sequestration in the Mitigation of Climate Change. Rome: FAO. 2010

[54] Young RR, Wilson B, Harden S, Bernardi A. Accumulation of soil carbon under zero tillage cropping and perennial vegetation on the Liverpool Plains, eastern Australia. Australian Journal of Soil Research. 2009;**47**:273-285

[55] Kelly RH, Burke IC, Lauenroth WK. Soil organic matter and nutrient availability responses to reduced plant inputs in shortgrass steppe. Ecology. 1996;77:2516-2527

[56] Hiltbrunner D et al. Cattle trampling alters soil properties and changes soil microbial communities in a Swiss sub-alpine pasture. Geoderma. 2012;**170**:369-377

[57] Schuman GE, Janzen HH, Herrick JE. Soil carbon dynamics and potential carbon sequestration by rangelands. Environmental Pollution. 2002;**116**:391-396

[58] Albaladejo J et al. Land use and climate change impacts on soil organic carbon stocks in semi-arid Spain. Journal of Soils and Sediments. 2013;**13**:265-277

[59] Raiesi F, Asadi E. Soil microbial activity and litter turnover in native grazed and ungrazed rangelands in semiarid ecosystem. Biology and Fertility of Soils. 2006;**43**:76-82

[60] Khalil MI, Haque MA, Sattar MA, Schmidhalter U. Relative contribution of crop residue bound-N to irrigated rice and carbon storage in a subtropical soil. In: Bernal et al., editors. Proceedings of the 11th International Conference of the FAO ESCORENA Network on Recycling of Agricultural, Municipal and Industrial Residues in Agriculture; Murcia, Spain; 2004. pp. 169-172

[61] Emadi M, Baghernejad M, Fathi H, Saffari M. Effect of land use change on selected soil physical and chemical properties in north highlands of Iran. Journal of Applied Sciences. 2008;**8**:496-502

[62] Gibon A. Managing grassland for production, the environment and the landscape. Challenges at the farm and the landscape level. Livestock Production Science. 2005;**96**:11-31

[63] MacDonald D et al. Agricultural abandonment in mountain areas of Europe: Environmental consequences and policy response. Journal of Environmental Management. 2000;**59**:47-69

[64] Dumont B et al. Rôles, Impacts et Services Issus des Élevages en Europe. Synthèse de L'expertise Scientifique Collective. France: INRA; 2016. p. 127. auto-saisine

[65] Pereira P, Brevik E, Muñoz-Miriam M, Miller B. Soil Mapping and Process Modeling for Sustainable Land Use Management. Elsevier; 2017. p. 398. eBook ISBN: 9780128052013

[66] Allen DE et al. What determines soil organic carbon stocks in the grazing lands of north-eastern Australia? Soil Research. 2014;**51**:695-706

[67] Molnár ZS et al. Common and conflicting objectives and practices of

herders and conservation managers: The need for a conservation herder. Ecosystem Health and Sustainability. 2016;**2**:e01215

[68] Dezsény Z, Drexler D. Organic agriculture in Hungary. Ecology and Farming. 2012;**3**:20-23

[69] Peco B et al. Effects of grazing abandonment on soil multifunctionality: The role of plant functional traits. Agriculture, Ecosystems and Environment. 2017;**249**:215-225

[70] Petz K et al. Mapping and modelling trade-offs and synergies between grazing intensity and ecosystem services in rangelands using global-scale datasets and models. Global Environmental Change. 2014;**29**:223-234

[71] Soussana J-F, Tallec T, Blanfort V. Mitigating the greenhouse gas balance of ruminant production systems through carbon sequestration in grasslands. Animal. 2010;**4**:334-350

[72] Khalil MI, Osborne B. Improving estimates of soil organic carbon (SOC) stocks and their long-term temporal changes in agricultural soils in Ireland. Geoderma. 2018;**322**:172-183

[73] Daryanto S, Eldridge DJ, Throop HL. Managing semi-arid woodlands for carbon storage: Grazing and shrub effects on above- and belowground carbon. Agriculture, Ecosystems and Environment. 2013;**169**:1-11

[74] Aubault H et al. Grazing impacts on the susceptibility of rangelands to wind erosion: The effects of stocking rate, stocking strategy and land condition. Aeolian Research. 2015;**17**:89-99

Chapter 5

Leaf Litter Decomposition and Mitigation of CO_2 Emissions in Cocoa Ecosystems

Askia M. Mohammed, James S. Robinson, David J. Midmore and Anne Verhoef

Abstract

Studies simultaneously quantifying litter weight losses and rates of CO_2-C evolved are few, though essential for accurate estimates of forest carbon budgets. A 120-day dry matter loss and a 130-day carbon emission experiments were concurrently conducted at the soil laboratory of the University of Reading, UK. Leaf litters of tree species comprising cocoa (*Theobroma cacao*), *Newbouldia laevis* (dominant shade tree in Eastern region (ER)) and *Persea americana* (dominant shade tree in Western region (WR)) of Ghana were incubated using a single tree leaf litter and/or a 1:1 mixed species leaf litters to determine and predict the litter decomposition and C dynamics in cocoa systems with or without the shade trees. Decomposition and C release trends in the ER systems followed: shade > mixed cocoa-shade = predicted mixed litter > cocoa; and in the WR, the order was: cocoa = mixed cocoa-shade > predicted mixed > shade. Differences between released C estimated from litter weight loss and CO_2-C evolution measurement methods were not consistent. Regression analysis revealed a strong (R^2 = 0.71) relationship between loss of litter C and the CO_2-C evolution during litter decomposition. The large C pool for shaded cocoa systems indicates the potential to store more C and thus, its promotion could play a significant role in atmospheric CO_2 mitigations.

Keywords: cocoa system, mineralization, mineralizable C, oxidizable C

1. Introduction

Soil organic matter is the main source of plant nutrients in low input agriculture [1], whilst the primary regenerative source of soil organic matter on agricultural lands is the decomposition of retained plant residues. Therefore, sustainable agricultural production based on nutrient cycling would operate only in systems where enough plant biomass is generated and retained on agricultural land. Hence, the success of forest ecosystems lies on their ability to store large amounts of organic matter aboveground in woody plant tissue and fibrous litter. Conceivably, biomass production, leaf litter decomposition and root biomass turnover in forest ecosystems have much influence on agro ecosystems' nutrient cycling and sustainability [2].

Nutrients are returned to soil through leaf falls and decomposition processes. Thus, their nutrient cycling starts during litter decomposition, where organically bound nutrients are released as free ions to the soil solution that then become available for uptake by plants. As the bulk of cocoa farmers are poor and therefore, do not apply external fertilizers to their croplands because of its high cost, litter decomposition from both cocoa and non-cocoa (or shade) trees plays a central role in the nutrition of the system. Although cocoa systems accumulate less leaf biomass than native forests, several studies have reported sizeable leaf litter-falls per hectare in cocoa systems (**Table 1**). The high biomass of leaf litter in cocoa systems indicates a high potential source of nutrient cycling when retained to undergo decomposition on the floor of the ecosystem but also a lot of CO_2 emissions.

Land use change is arguably the most common anthropogenic activity that interferes greatly with most biogeochemical cycles. Its impact on C and nutrient cycling in the soil has been the subject of much attention [3–6]. Understanding the effects of land use/land cover change on ecosystem functions is often derived by quantifying changes in C and nutrient stocks and fluxes. Indeed, changes in forest cover have been implicated in the rising levels of carbon dioxide (CO_2), the main greenhouse gas (GHG), in the atmosphere [4, 7, 8]. This is because large amounts of organic C are often stored in forest trees and which upon clear-felling, decompose and release the stored C to the atmosphere [9].

Since plant litter decomposition involves carbon dioxide (CO_2) emissions, and fragmentation and leaching of organic matter to the soil, many studies have been conducted to investigate the factors controlling litter decomposition. These studies have often followed one of three approaches. The first approach simply measures the annual litter-fall in vegetation and equates that to the amount of organic material being decomposed. This approach acknowledges that the soil organic matter level in most natural vegetation types attains an equilibrium state where the amount of material being decomposed annually is equal to the amount of annual litter-fall. The second approach is the weight loss of buried litter over time, whilst the third approach measures the microbial activity on litter via CO_2 evolution.

Using weight loss of buried litter contained in nylon mesh bags over time, many studies concluded that the rate of leaf litter decomposition depended on tree species, the chemical composition of the leaves, and environmental factors such as temperature and soil moisture [2, 10, 11]. Hitherto, several researchers considered the C/N ratio of plants as the main plant composition factor that controls decomposition rates [10]. Increasingly, other litter constituents such as lignin and polyphenol concentration, especially in the tropics, are considered to play important roles in the decomposition process [11–14].

Country	Annual litter-fall				
	Age cocoa (years)	Cocoa tree	Shade tree	Total	Reference
Malaysia	8–10	5460	2660	8120	[17]
Venezuela	30	7630	13,571	21,201	[18]
Costa Rica[a]	10	nd[b]	nd	7071	[19]
Costa Rica[c]	10	nd	nd	8906	[19]
Brazil	nd	nd	nd	9000–14,000	[20]
Ghana	nd	nd	nd	5000	[21]

[a]*Cocoa with* Cordia alliodora *as shade tree.*
[b]*No data.*
[c]*Cocoa with* Erythrina poeppigiana *as shade tree.*
Source: from [22].

Table 1.
Annual litter-fall of cocoa ecosystems (in kg dry matter (DM)/ha).

Although the use of litter bags makes it possible for buried litter recovery, their use for decomposition studies has been criticized for creating unrealistic 'microclimate' conditions; e.g. moisture may be elevated to levels not found in unconfined conditions and can contribute to litter decomposition [15, 16]. Losses due to earthworm consumption or litter fragmentation overestimate those mediated by microbial activities. Frankland [23] studied the weight loss pattern of *Bracken rachids* through the unconfined method i.e., using plastic labels to mark the litter before placing it on the soil surface. The study noted that retrieval of decomposing litter under this method is tedious and presents large potential errors in estimating the decomposition rates. Also, it does not protect the litter from the interferences of larger organisms and fragmentation. Therefore methods to overcome these drawbacks are imperative when investigating the factors that control leaf litter decomposition.

As an alternative, numerous studies have measured decomposition rates of litter by the CO_2 evolution or carbon mineralization method [14, 24–29]. This method has the advantage of measuring decomposition during shorter periods (hour scale); i.e., through early decomposition stages when weight loss cannot be accurately quantified [30]. However, the CO_2 evolution method requires an experimental set-up that often deviates farther from the natural conditions than weight loss measurement. Studies simultaneously quantifying litter weight losses and investigating rates of CO_2-C evolved are few; however they are essential for accurate estimates of forest nutrient cycling. Nonetheless, results from the few studies on litter decomposition studies using the CO_2 releasing method have been comparable to the weight loss method [27, 31, 32].

Many litter decomposition studies have focused on individual plant species when investigating the factors influencing litter decay [33–38]. However, leaf litters in ecosystems with more than one dominant plant species do not fall separately, either in time or space, but create an admixture of litters. Although the potential of litter interactions was hypothesized to have a marked effects on their decomposition in agro ecosystems many years ago by Thomas [39], Staaf [40], Seastedt [41] and many others, studies on potential interactions in mixed leaf litter decomposition are still few and not well understood, and so require further investigations to aid planning for nutrient management through decomposition and nutrient release in agroforestry and similar systems [42]. Hansen and Coleman [43] noted some changes in the chemical environment (increased nutrient availability) due to litter mixing during decomposition studies of mixed litters of yellow birch (*Betula alleganiensis* Britton), red oak (*Quercus rubra* L.) and sugar maple (*Acer saccharum* Marsh).

This chapter reports on the findings from laboratory incubation experiments carried out on separate leaf samples of cocoa and shade species and 1:1 mixed cocoa-shade leaf litters. Dry mass loss and C emission from unconfined leaf samples were analyzed to (i) determine the decomposition dynamics of cocoa litter, and of the dominant shade species in cocoa ecosystems, (ii) investigate the effects of leaf interactions on decomposition (iii) assess the relationship between decomposition rates and C release patterns, (iv) determine the C emission rates during leaf decomposition, and (v) assess the relationship between leaf weight loss and C emissions during decomposition, and (vi) determine the CO_2 mitigation potentials of cocoa systems. The study hypothesized that the decomposition rates of the mixed leaves of cocoa-shade systems would (1) differ from the rates of decomposition of the separate litter components decomposing alone (i.e., separate cocoa and shade leaves), (2) be equal to the pooled rates of the separate litter components decomposing alone, (3) the amount of C in the litter loss is the same as the C emitted during litter decomposition and (4) cocoa systems have the potential to mitigate CO_2 emissions.

2. Materials and methods

2.1 Leaf sampling and site

Leaf samples from three tree species (cocoa, *Newbouldia laevis* dominant species in Eastern Region (ER) and *Persea americana* dominant in Western Region (WR)) were collected from cocoa ecosystems in ER and WR of Ghana. Farms were selected in the ER at the Duodukrom community within the Suhum District (6°2′ N, 0°27′ W), and the in WR at Anyinabrim found in the Sefwi-Wiawso District (6°57′ N, 2°35′ W).

The ER covers a land area of 19,323 km^2 representing 8.1% of the total land area of Ghana [44]. It lies between latitude 6° and 7° N and longitude 1°30′ W and 0°30′ E. The region has been producing cocoa long before cultivations started in the Western region. The WR occupies a land area of 23,921 km^2 which is approximately 10% of the total land area of Ghana [44]. The region is the wettest part of Ghana and harbors about 24 forest reserves that account for about 40% of the forest reserves in Ghana. The sampled leaf litters from these regions were transported to the Soil Research Centre of the University of Reading, UK, where the following experiments were conducted.

2.2 Initial chemical properties of the oven-dried leaf litters

2.2.1 Total carbon

The total carbon (C) in the samples was determined using the Europa Roboprep connected to a VG 622 Mass Spectrophotometer. Weights of 0.90–1.10 mg (oven dry) of plant components (root, stem, branch, leaf and litter), and 8.00–12.00 mg (air dry) of soil samples, in triplicate, were put into small pre-weighed aluminum cups and pressed to seal completely using forceps. The sealed samples were arranged in a labeled sample holder and transferred to the Mass Spectrometry System for analysis. The analytical output was in % C of the samples.

2.2.2 Total nitrogen

The total N in the leaf and soil samples was determined using the Europa Scientific ANCA System. Samples of 5–6 mg leaf and 8–12 mg soil were weighed into small aluminum cups and pressed to seal using forceps. The sealed samples were transferred to the Europa system for analysis. The analyzed data were expressed as % N (w/w).

2.2.3 Total P, K, Ca, Mg and S

Approximately 0.5 g oven dry plant samples (i.e., root, stem, branch, leaf) were accurately weighed and transferred into 100 mL Kjeldahl digestion tubes. About 10 mL of concentrated AnalaR nitric acid were added to each tube in a fume hood. Each tube was then covered with a glass marble and left to stand overnight. The tubes were placed on a digestion block the next day and cautiously heated to 60°C for 3 h followed by a gradual increase to 110°C and allowed to digest for 6 h. The digestion tubes were then removed, allowed to cool and the digest filtered through prewashed No. 540 (12.5 cm diameter) filter papers into 100 mL volumetric flasks. The flasks were made up to volume with ultra-pure water. Aliquots of 5 mL from each flask were diluted by a factor of two and analyzed for concentrations of P, K, Ca, Mg and S using the inductively coupled plasma-optical emission spectrometry (ICP-OES).

Standards of multi-element solution (0.5, 1, 50 and 100 mg/L K, Ca, Mg, Mn, Zn, Fe, and Al), sulfur (50 mg/L) and phosphorus (50 mg/L), as well as a blank (0 mg/L) were prepared to contain the same nitric acid concentration as in the digest to calibrate the ICP-OES. The data generated by the ICP-OES were reported in concentrations (μg/L) which, after correcting for the blank reading, was converted to mg/kg dry weight based on the sample weights digested, volume of extract and the dilution factor [45].

2.2.4 Lignin concentration

As outlined in Anderson and Ingram [46], a 1 ± 0.001 g sample of leaf for each tree species was weighed (W1) into 200 mL Berzelius beaker. A 100 mL of acid detergent solution (20 g of cetyltrimethyl ammonium bromide (CTAB) was dissolved in 27.84 mL of sulfuric acid (98% purity) in a 1000 mL volumetric flask and brought to the mark with distilled water and to form a clear solution by heating) was then added and heated to boil for 1 h. The content was filtered hot through a vitreosil crucible (No. 1) of known weights (W2). The residue was washed with 3 × 50 mL aliquots of hot water and then with acetone until no more color was removed. The residue was then oven-dried at 105°C for 2 h, cooled in a desiccator and weighed whilst still in the crucible (W3). The sample remaining expressed as a percentage of the initial weight of the sample, estimated the acid detergent fiber (ADF) content of the sample:

$$\text{ADF}(\%) = \frac{(\text{W3} - \text{W2}) \times 100}{\text{W1}} \tag{1}$$

A saturated potassium permanganate solution was prepared by dissolving 50 g $KMnO_4$ and 0.05 g Ag_2SO_4 in a 1000 mL volumetric flask and brought to the mark with distilled water. Lignin buffer solution was also prepared by dissolving 6 g $Fe(NO_3)_3{\cdot}9H_2O$ and 0.15 g $AgNO_3$ in water followed by addition of 400 mL methylpropan-2-ol and diluted to 1000 mL with distilled water. A combined solution of the saturated $KMnO_4$ and lignin buffer solution in the ratio of 2:1 was prepared. The crucible containing the ADF was then placed in a shallow enamel containing cold water carefully without wetting the fiber and 25 mL of the combined $KMnO_4$/buffer added. The content was stirred with a glass rod to break up lumps and to wet all the fiber particles in the crucible with the solution and allowed to stand for 3 h. The content in the crucible was then filtered under suction and washed with demineralizing solution (50 g oxalic acid dehydrate dissolved in 700 mL 95% ethanol, followed with addition of 50 mL conc. HCl and diluted to 1000 mL with distilled water) until white. This was filtered and washed thoroughly with ethanol under continuous suction and washed in a similar manner with acetone. The crucible was then oven-dried at 105°C for 2 h, cooled in a desiccator and weighed (W4). The percentage lignin in the sample was then calculated as:

$$Lignin(\%) = \frac{(W3 - W4) \times 100}{W1} \tag{2}$$

2.3 Sample preparation for experimentation

Approximately 100 g each of 2-mm sieved air-dried plant materials (viz. cocoa, *N. laevis* and *P. Americana,* 1:1 (w/w) mixture of cocoa: *N. laevis* and cocoa:

P. americana leaf litters) were weighed into a 500 mL beaker. Water (300 mL) was gradually added to each weighed litter sample with continuous stirring to produce a moist litter treatment that was not saturated [27]. They were kept in a fridge at a temperature of 4°C for 72 h to attain an equilibrated moisture status in all treatments. Three weighed sub samples (6 g) of each treatment were oven-dried at 80°C for 24 h and used to determine a conversion factor to an oven dry basis.

2.4 Leaf decomposition experiment

A known weight (approx. 6 g) of each leaf treatment was transferred into a labeled 15 mL beaker separately. Soil (~5 mg) from the region specific to the leaf treatment was added to the beaker to serve as an inoculant. Each beaker unit was replicated 12 times to give the total of 72 experimental units. These units were weighed and randomly arranged for incubation in a 30°C controlled dark room located in the Soil Chemistry laboratory of the Soil Research Centre, University of Reading, UK. Three (3) replicates of each treatment were retrieved after 0, 20, 50, 80 and 120 days of incubation. The beakers so retrieved following each incubation period, were oven-dried at 80°C for 24 h and weighed. The residual oven-dried litters were appropriately labeled and stored for chemical analysis.

2.5 CO_2-C emission experiment

A sample (3 g) of each litter treatment was transferred into a 250 mL conical flask. As shown in **Figure 1**, the neck of the flask was closed with a rubber bung from which was suspended a vial containing 20 mL of 1 M NaOH solution to trap CO_2 evolved as outlined by Rowell [47]. A similar conical flask was set-up without leaf treatment as a blank. Each flask unit was replicated 3 times to give a total of 21 experimental units comprising 18 litter treatments and 3 blanks. Also, the treated conical flasks were randomly arranged for incubation in a 30°C controlled dark room located in the Soil Chemistry Laboratory of Soil Research Centre, University of Reading, UK. At 0, 3, 5, 11, 16, 28, 43, 60, 75, 90, 103 and 130 days of incubation, the vials were removed, and the NaOH was carefully transferred quantitatively (with rinsing) into an empty 50 mL conical flask for titration. Ten milliliters (10 mL) of 1 M $BaCl_2$ was added to precipitate the carbonate compounds ($NaHCO_3$) formed as a result of reaction between NaOH and CO_2 (**Figure 1**). The vials were thus, removed 12 times and replaced after refilling with fresh NaOH solution before closing the incubation beaker to continue the capture of released CO_2 from the decomposing leaves. The amount of CO_2 captured was determined by titrating the unreacted NaOH in the 50 mL flask with 0.5 M HCl using phenolphthalein as the indicator.

2.6 Data analysis

The data on per cent mass remaining, carbon and nutrient concentrations of pure cocoa leaf and shade species leaf decomposing alone were used to estimate expected data for mixed cocoa and shade litter denoted as predicted mixture, using the simplified form of similar relations used by others as:

$$predicted(x) = \frac{Cocoa(x) + Shade(x)}{2} \qquad (3)$$

where x = per cent mass remaining, carbon or nutrient concentration, and C emission of the leaf treatments at each retrieval day [28, 48, 49]. Any significant difference between the estimated predicted mixture value and the actual mixed

Figure 1.
Photograph of CO_2-C emission experiment.

cocoa-shade leaves treatment indicated an interaction in the decomposition of the mixed leaves, either negative or positive [49]. The data on % residual leaf were fitted to the exponential decay Eq. (2) that was proposed first by Olson [50] to describe the decomposition rates of the leaf litter treatments:

$$R_t = R_o * \exp(-k_d t) \tag{4}$$

where R_t = % residual weight at time t, R_o= initial litter per cent at day zero (i.e 100%) and k_d = decomposition rate constant.

The data on C emission were fitted to the single exponential rise-to-maximum (growth) model.

$$C_t = C_o * [1 - \exp(-k_m t)] \tag{5}$$

where C_t = amount of C emitted after time t of the incubation; C_o= amount of C that can be potentially emitted within the period of incubation; and k_m = mineralization rate constant.

The amounts of C accompanying the loss leaves during the decomposition processes were calculated as:

$$[C]_{loss(t)} = \frac{(100 - \%leaf_{remained(t)})}{100} \times [C]_i \tag{6}$$

where $[C]_{\text{loss(t)}}$ is amount of C in the leaf loss (g/kg) at time t (i.e., 0, 20 40, 80 and 120 days), and $[C]_{\text{i}}$ is the initial carbon concentration in the litter (g/kg).

The comparison was carried out statistically using ANOVA to test for significant differences of all data parameters (% residual litter mass, C and nutrient concentrations and release, C emission, C_o, and k_d). Tukey's mean separation procedure at the 0.05 level of significance was used for all data. All figures were produced with SigmaPlot 10.0 using the means of % residual litter mass, C and nutrient concentrations and release and also C emission. Also the fitted model parameters were estimated using the SigmaPlot 10.0 regression analysis module.

3. Results and discussion

3.1 Chemical characteristics of the leaf sample

3.1.1 Eastern region of Ghana

Literature is replete with the important role that litter chemistry plays in decomposition and nutrient release in top soils [14, 26, 51–53]. The initial concentrations of some elemental nutrients of the leaf litter samples are presented in **Table 2**. The leaf litter treatments did not differ significantly ($P > 0.05$) in their C and S concentrations in the Eastern region (ER). The leaf litters generally had low oxidizable carbon concentration by weight with less variability as indicated by the narrow range of 337.0–382.6 g/kg. The cocoa-shade mixture had the highest C and the leaf litter of shade tree contained the least (**Table 2**). Honeycutt et al. [54], Woomer et al. [55], and Kuo et al. [56] noted that plant residues have stable carbon concentration of about 40% by weight. Other studies have used a constant value from 450 to 500 g/kg as the C concentration for all parts of a tree biomass including the leaves [57–60]. Thus, the present C concentrations of the leaf litters from cocoa ecosystems in the ER of Ghana are lower than the commonly used ranges. The C concentration of 357.2 g/kg for the leaf litter of cocoa is lower than that reported by Anglaaere [16] as 413 g/kg for cocoa leaf samples from Atwima District in the same Eastern region of Ghana where the present study was carried out. However, the present litter C ranged confirmed the reported C data of 364 to 400 g/kg on similar cocoa leaf litter by Ofori-Frimpong and Rowell [27].

Among the nutrients that varied significantly between the leaf samples, that of the shade tree *Newbouldia laevis*, had significantly ($P < 0.05$) higher K, Ca and lignin but lower Mg concentration than the cocoa leaf litter (**Table 2**). The N, P and S concentrations in the leaf litters of the cocoa and shade trees did not differ significantly ($P > 0.05$). The predicted mixture contained higher P and Mg concentrations than

Leaf litter	C	N	P	K	Ca	Mg	S	L	C/N
					g/kg				ratio
Eastern region									
Cocoa	357.2a	18.4ab	1.13a	4.58c	12.9c	5.94a	1.89a	215.7c	19.4a
Shade	337.0a	13.8b	1.07a	7.36a	16.3a	3.00d	1.89a	288.0a	25.2a
Coc-shade	382.6a	19.5a	0.99b	5.61b	14.1b	4.12c	1.83a	256.9b	19.6a
Predicted	347.1a	16.1b	1.10a	5.97b	14.6b	4.47b	1.89a	251.8b	22.3a
Western region									
Cocoa	378.7a	17.4a	0.85a	8.57a	11.0a	5.25a	1.65b	220.0c	21.9a
Shade	342.4a	15.2a	0.79a	4.43b	9.1b	2.67b	1.78a	289.1a	23.6a
Coc-shade	360.2a	18.5a	0.85a	6.55c	10.0c	3.96c	1.72ab	265.4b	19.5a
Predicted	360.6a	16.3a	0.82a	6.50b	10.0b	3.96b	1.72ab	254.5b	22.7a

[1]Newbouldia laevis *in Eastern region,* Persea americana *in Western region.*
[2]*Different letters within same region and column indicate significant difference at* $P < 0.05$ *using Tukey's method.*

Table 2.
*Initial chemical composition and lignin (L) content (mean) and nutrient ratios of the leaf litters of cocoa, shade (*Newbouldia laevis, Persea americana*)[1] and mixed cocoa-shade under cocoa ecosystems in eastern and Western regions and the predicted mixed litter in Ghana.*[2]

the analyzed mixture of cocoa-shade leaf litter. The lower P and Mg concentrations in mixed cocoa-shade litter than expected or predicted is an indication of non-additive response or negative interaction when cocoa and shade leaves are mixed together. The other nutrient elements, however, were similar in the mixed cocoa-shade and predicted mixed litters, indicating an additive response for those nutrient elements in mixed litter systems such as shaded or agroforestry systems (**Table 2**).

The initial N concentrations of all the leaf litters were higher than the critical level for cocoa foliage N concentration (9.0 g/kg), below which point net N immobilization would be expected [61]. Thus, with the high N concentration of the leaves, net N mineralization is highly possible during the leaf decomposition. Several researchers have considered the N and/or its ratios such as C/N, lignin/N and polyphenol/N of residues as major factors controlling decomposition processes [14, 34, 52]. The C/N and lignin/N ratios of the leaf litters from ER ranged from 19.4 to 25.2 and 11.7 to 21.8, respectively (**Table 2**). Thus, the C/N ratios of the litters are so close to the critical level of 25 noted for decomposition and N mineralization to occur [62, 63].

Where nutrient ratios are used as indices of nutrient status to microbial growth, Girisha et al. [64] put forward that nutrient retention during decomposition depends on their initial status in the litter. The C-element ratio has been commonly used to explain nutrient status where a nutrient element, *say* R, is said to be limiting when the C/R ratio is above a certain critical level set for microbial growth. In this case, nutrient element R will be retained resulting in immobilization but where C/R ratio is below the critical level, the nutrient element R is released during decomposition of the litter [65]. The C/N ratios were statistically similar in the leaf litter treatments under ER (**Table 2**).

Table 2 revealed that the P, K and Mg concentrations of all the leaves from the ER were less than the critical range of 2.0 to 2.5 g/kg, 20 and 5 g/kg respectively. As such, P, K and Mg immobilization would be expected during decomposition [61, 66–68]. The Ca concentrations of the litters were higher than the critical 6 g/kg value. The lignin content of the leaf litters from the ER ranged from 220.0 to 289.1 g/kg dry matter (DM). The cocoa leaf litter had significantly ($P < 0.05$) lower lignin status when compared with the shade and the mixed cocoa-shade leaf litters, as well as the predicted mixture (**Table 2**). However, the lignin concentration (215.7 g/kg DM) in the cocoa leaf litter in this study is higher than the data (141–146 g/kg DM) of Dawoe [69] on cocoa leaf lignin status in Ghana. The lignin concentration in mixed cocoa-shade litter was similar to the predicted mixed litters (**Table 2**). Lignin has been considered a determinant of litter quality and a predictor of decomposition by previous researchers [34, 70, 71].

3.1.2 Western region of Ghana

The leaf litters from the Western region (WR) varied considerably in their initial nutrient and lignin concentrations (**Table 2**). The variations in the C, N and P concentrations did not however, differ significantly ($P > 0.05$) among the leaf litter treatments (**Table 2**). The C concentration of cocoa leaves in the present study was lower than the reported data by Anglaaere [16]. This study found the N (1.74%) in leaf litter of cocoa to corroborate previous studies in other parts of Ghana [16, 69]. Anglaaere [16] reported N of 18.5 g/kg for cocoa leaves, whilst Dawoe [69] found an N range of 9.5 to 14.8 g/kg for cocoa leaves from 3 to 30 year old cocoa trees. It thus appears that foliar N concentration for cocoa is highly variable. With the N concentration of all the leaf litters being lower than the critical N level of 20 g/kg, net N immobilization would be expected during decomposition [29, 62, 63].

Differences in total P concentration of the leaf litters from the WR were not significant ($P > 0.05$) and were all low when compared to the leaf litters from ER

(**Table 2**). Anglaaere [16] also found the P to be low, at 1.3 g/kg in Ghanaian cocoa leaf, confirming the low P status of cocoa litters in Ghana found in the two regions by this study. Compared with critical levels of nutrient elements, the cocoa leaf litter in this study is deficient in P, K and Mg concentrations [61]. It is thus expected that during decomposition of the leaf litters from WR, as with ER, the deficient elements would not be released easily resulting in temporal nutrient immobilization [61, 66, 67].

The K, Ca and Mg concentrations in the leaf litter of the WR shade tree *P. americana* were significantly ($P < 0.05$) lower than those in the cocoa litter but higher in lignin concentration (**Table 2**). With respect to K and Mg, the leaf litter of the cocoa in the WR contained approximately twice as much as that in the shade tree litter. Compared with data on cocoa foliar nutrients in Anglaaere [16], the cocoa leaf nutrients from the WR appeared to be similar.

The initial chemical composition of the leaf litter of mixed cocoa-shade was generally similar to that of the predicted mixture treatment with the exception of P and Mg in the ER; and K, Ca and Mg in the WR, suggesting a high predictability of mixed litter nutrients from the single component species nutrient concentrations (**Table 2**). Overall, there were significant variations in the nutrient balance of the leaf litters and also high variability in nutrient ratios as shown in **Table 2**. These nutrient variations were more pronounced in the single litter treatments than the mixed and predicted mixed litters.

3.2 Decomposition trends of the leaf samples

Figure 2a presents the decomposition patterns of leaf litters of cocoa ecosystems in the ER obtained during a 120-day laboratory incubation experiment. During the first 20 days of incubation, the shade litter lost approximately 9% of its initial weight whereas the cocoa leaf litter lost only 2.8% of its weight, indicating a lag phase in the decomposition of the cocoa leaf litter (**Figure 2a**). Anglaaere [16] reported mass loss of 3.45% of cocoa leaf litter within the first one month of initial decomposition. However, the % leaf litter of cocoa and shade trees that remained did not differ significantly during the first 20 days of incubation. Thereafter, significant differences occurred in the per cent mass remaining between the leaf litters of cocoa and shade trees, with the leaf litter of the latter continuously decomposing at a higher rate than the former as incubation time progressed (**Figure 2a**). At the end of the 120 days of incubation, the leaf litter losses were 17.6 and 30.7% in the cocoa and shade litters, respectively. These litter loss percentages compared well with the reported losses between 16 and 33% of cocoa leaf litters within 80 days of incubation by Ofori-Frimpong and Rowell [27].

Overall, the decomposition pattern of the mixed cocoa-shade litter treatment indicated an additive response and thus, appeared predictable from the decomposition patterns of the component litters decomposing alone. Although the litter remains of the mixed cocoa-shade litter could not be separated into the individual components at any stage of the incubation, both the predicted and the actual mixed cocoa-shade litter treatments indicated higher ($P < 0.05$) decomposition rates than the pure cocoa leaf litter treatment (**Figure 2a**); this indicates that mixing leaf litter has the potential to enhance litter decomposition in the cocoa ecosystems in the ER.

The decomposition patterns of the leaf litter gathered from the cocoa farms in the WR is presented in **Figure 2b**. The decomposition pattern of the leaf litter of the shade species (*P. americana*) displayed a lag phase within the first 20 days of incubation where only about 3.6% had undergone decomposition (**Figure 2b**). The leaf litter of cocoa had significantly lower % remaining when compared with the

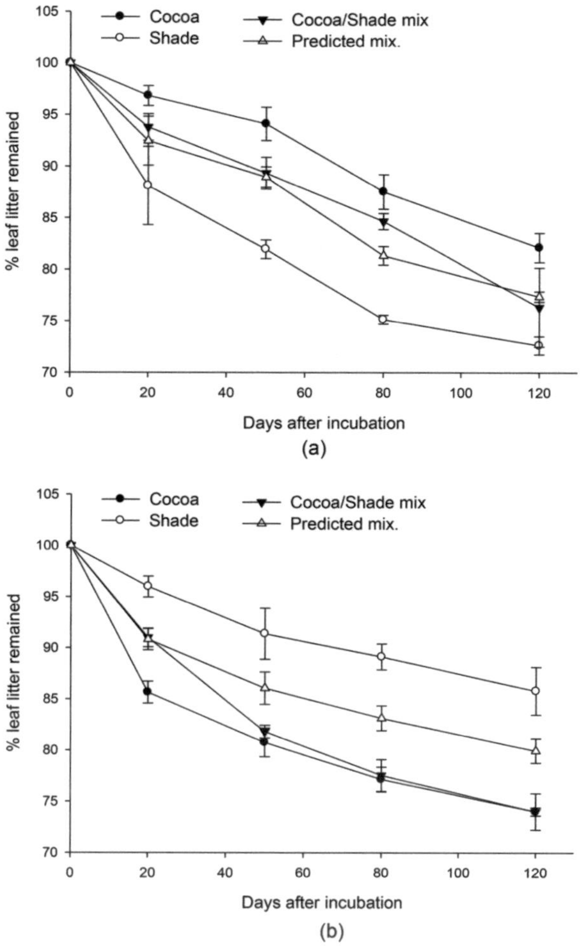

Figure 2.
Decomposition patterns of leaf litters in cocoa ecosystems: (a) Eastern region and (b) Western region. Bars indicate standard error (n = 3).

leaf litter decomposition pattern of the shade species as the incubation progressed. By the end of 120 days of incubation, the mass loss of cocoa was 29.1% of the initial weight whilst the shade litter decomposed by only 15.1% of its initial dry weight (**Figure 2b**). The mass loss of 29.1% for cocoa leaf litter from WR is greater than that (17.6%) from ER and, also, the reported mass loss of 22.35% by the 4th month of incubation of cocoa leaf litter from the Ashanti region of Ghana [16].

Comparison between the decomposition patterns of mixed cocoa-shade and the predicted mixed litter treatments showed no significant ($P > 0.05$) difference up to the first 20 days of incubation. Thereafter, there was a clear difference in the decomposition patterns between them for the next 100 days of incubation with the mixed cocoa-shade litter treatment decomposing at faster rates. As the time for decomposition progressed beyond 50 days, the cocoa-shade mixed litter decomposition pattern tended to behave like the pure cocoa leaf litter (**Figure 2b**). Thus, the mixed cocoa-shade treatment showed a non-additive response with positive interaction after 20 days of incubation and therefore, its decomposition is not predictable from the component species of the mixture. Over the 120-day incubation period, the amounts of leaf litter that decomposed were 26 and 20% for the mixed cocoa-shade litter and predicted mixed litter, respectively. The implication is that shaded cocoa system in the WR would be more efficient in nutrient cycling than expected. However, decomposition of cocoa leaf litter alone (unshaded system) was similar to that of mixed cocoa-shade litter (shaded systems).

3.2.1 Decay constant

The leaf decomposition patterns shown in **Figure 2a** and **b**, conformed well (R^2 = 0.81–0.99) to the single exponential decay model proposed by Olson [50] for similar mass loss studies. However, the plots of the model were observed not to go through 100% weight remaining at the start (i.e., day zero) of the incubation, indicating some degree of deficiency associated with the model [2]. Notwithstanding, the same model has been used by other researchers in recent times to fit similar litter decomposition data [46, 72–74]. From the fitted model for the % leaf litter remaining, the decomposition rate constants (k_d) of the leaf litter treatments were estimated for a period of 120 days of decomposition and are presented in **Table 3**. The k_d measures the proportion of the material that decays per unit time. Therefore, the lower the k_d of a decomposing organic material, the slower it decomposes.

In the Eastern region (ER), the decomposition rate constants varied considerably and ranged from 1.65×10^{-3}/day for the leaf litter of cocoa to 2.72×10^{-3}/day for leaf litter of shade tree (**Table 3**). The k_d for the mixed cocoa-shade and predicted mixture treatments were intermediate between that of the cocoa and the shade, in support of their decomposition patterns described earlier (**Figure 2a**). The leaf litter of the shade species had a much higher ($P < 0.05$) decay rate constant than the cocoa leaf litter treatment. Anim-Kwapong and Osei-Bonsu [75] found a similar high k_d value of 2.51×10^{-3}/day (recalculated from authors' half-life value of 9.22 for *N. laevis*) for *N. laevis* also collected from the same ER of Ghana. In respect of the decomposition rate of cocoa leaf litter in the present study, comparable leaf litter decay rates have been reported by Owusu-Sekyere et al. [76] and for some forest tree species [77]. The decomposition rate of the predicted mixed leaves and the actual mixed cocoa-shade treatments (2.14×10^{-3}/day) were exactly the same; this agrees with the earlier assertion of additive response for mixed leaf decomposition

Leaf litter treatment	Decomposition constant, k_d/day	Potential C mineralizable, C_o g/kg
Eastern region		
Cocoa litter	0.00165b[2]	107.7b
Shade litter	0.00272a	114.8ab
Mixed cocoa-shade	0.00214ab	118.5a
Predicted mixed litter	0.00214ab	111.2ab
Western region		
Cocoa litter	0.00240a	119.9b
Shade litter	0.00127b	61.10d
Mixed cocoa-shade	0.00259a	209.7a
Predicted mixed litter	0.00187ab	90.40c

[1]Newbouldia laevis *in Eastern region,* Persea americana *in Western region.*
[2]*Different letters within same region and column indicate significant difference at* P < 0.05 *using Tukey's method.*

Table 3.
*Decomposition constants and potential mineralizable C (% of initial oxidizable C) of cocoa, shade (*Newbouldia laevis, Persea americana*)*[1]*, mixed cocoa-shade and predicted cocoa-shade mixture under cocoa ecosystems in eastern and Western regions of Ghana.*

simulating the litter state of shaded cocoa ecosystems in the ER. Based on the k_d values, the order of the leaf litter decomposition in the ER followed:

Shade > mixed cocoa-shade = predicted mixed litter > cocoa.

The higher rate of decomposition in mixed leaves of cocoa-shade than pure leaves of cocoa suggests that nutrient cycling in cocoa ecosystems of the ER would be favored by shaded cocoa ecosystems. Owusu-Sekyere et al. [76] attributed the slow decomposition rate constant of cocoa to high lignin and polyphenol concentrations in the cocoa leaves. However, in their case the data showed no significant difference between leaves of forest species and cocoa with respect to C/N ratios, yet the forest species exhibited a significantly higher decay rate.

Indeed, several works on litter decomposition have reported significant correlation between initial chemical composition of decomposing materials and the decay constants. Some of the litter constituents that have indicated significant correlations with k_d in previous studies included the initial C, N, P, Ca concentrations, lignin, polyphenols and ratios of C/N, lignin/N and polyphenol/N of decomposing organic materials [10–14, 32, 76]. In contrast, the k_d values estimated for the leaf litters from ER did not show significant correlations with C, N, P, S and C/N (**Table 4**). Rather, the k_d values correlated positively ($P < 0.05$) with K, Ca, and negatively ($P < 0.05$) with Mg (**Table 4**). The cations; K, Ca and Mg are known to activate enzymes that promote metabolism [78] implicating their role to driving litter decomposition. Similar correlations between decomposition rates and initial concentrations of K, Ca, and Mg have previously been observed by Briones and Ineson [31]. The lack of correlations between k_d values and N concentration or C/N ratios could partly be due to the narrow ranges of the C and N concentrations of the leaf litters under study and/or partly indicates the insensitivity of the k_d to assess the decomposability of the litters [79–82]. The current study agreed with McTiernan et al. [83] who found a significant correlation between k_d and Ca, and none for k_d and N concentration for a mixture of tree leaf litter during decomposition. Elsewhere, others also found C/P ratios to correlate with litter decomposition [84].

In the Western region (WR), decomposition rate constants of leaf litters under cocoa ecosystems ranged from 0.00127 to 0.00259/day (**Table 3**). The estimated k_d values for cocoa systems are higher than the k_d values ranging from 0.221 to 0.227/y (i.e., 0.00060 to 0.00062/day) in Dawoe [69] for cocoa systems in the Ashanti

Element	Eastern region	Western region
C	0.359	0.362
N	−0.549	0.301
P	−0.390	0.799**
K	0.766**	0.726**
Ca	0.726**	0.689*
Mg	−0.763**	0.715*
S	−0.392	−0.496
L	0.766**	−0.591*
C/N	0.664*	−0.195

**Correlation is significant at the 0.05 level (2-tailed).*
***Correlation is significant at the 0.01 level (2-tailed).*

Table 4.
Correlation coefficients (r) between initial chemical composition and lignin (L) of leaf litter and decay constant (kd) from cocoa ecosystems in the Eastern and Western regions of Ghana.

region of Ghana. The author used data on monthly litter fall to estimate the k_d values using the annual litter-fall over litter-stock formulae, which has a greater source of variability due to seasonal effects on monthly litter-fall. Depending on the season, particularly during wet seasons, less litter fall may be recorded leading to lower estimates for k_d values. Anglaaere [16] reported much lower litter-fall biomass from April to November which coincided with the wet season period in the cocoa cultivation regions of Ghana.

The decomposition of the cocoa leaf treatment was significantly faster than that of the shade species (**Table 3**). The difference in the decomposition rates of leaves from cocoa and the shade species is attributable to the differences in the biochemical composition of their leaf structures as similar attributions have been made by many workers to explain variations in decomposing organic materials [28, 64, 81, 82]. Indeed, the present study found significant ($P < 0.05$) positive correlations between k_d values and K, Ca, and Mg but negative correlations with C and S (**Table 4**). Briones and Ineson [31] also reported significant correlations between k_d and K, Ca, and Mg in decomposition of eucalyptus, ash and birch individually, and as litter mixtures during decomposition. Of the initial concentrations of chemical parameters, the cocoa leaf litter was significantly higher in all the positively correlated parameters and lower in all the negatively correlated parameters than the leaf litter of the shade tree, hence the higher estimate of its decomposition rate constant (**Table 3**).

The decomposition rate appeared faster in leaf litter treatments of mixed cocoa-shade than the predicted mixture from the single decomposition rates of the components, but the difference was not significant ($P > 0.05$, **Table 3**). The observed faster rate of decomposition in the mixed cocoa-shade is an indication of a possible synergistic effect under shaded cocoa systems in the WR during litter decomposition. Compared with the single plant species leaf litter decomposition rate constant, the mixed leaf litter decomposed at a rate similar to leaf litter of pure cocoa, both of which were faster ($P < 0.05$) than the pure leaf litter of the shade tree. The significance of the above finding is that, litter decomposition and nutrient release patterns would not differ between shaded cocoa and unshaded cocoa ecosystems in the WR but will be lower in a forest of only *P. americana* trees.

3.2.2 Carbon release patterns

The carbon and nutrient contents of the residual litters were determined as the product of their concentration and the litter dry mass; this allowed C and nutrient release to be plotted as a percentage of the initial C and nutrient contents of the litters. Similar plots have been provided by other researchers [48, 49]. **Figure 3** presents the C release patterns for the various litters during the course of decomposition. Leaf litters from the ER all released C in the course of the decomposition. The C release patterns were similar and linear except in the mixed cocoa-shade where the pattern was curvilinear with an initial faster C release within the first 20 days, then a gentle release between 20 to 80 days, and a slow release followed thereafter to 120 days (**Figure 3**ER: (C)). However, the amount of carbon released among the litters did not differ significantly ($F_{12, 40} = 0.64$, $P = 0.797$) during the decomposition. Released C during litter decomposition has been attributed to losses of soluble C and mineralization [65]. After 120 days of incubation, the C released from litters of ER varied between 23.3 and 26.8% of the initial litter C contents.

With regard to decomposition of litter from WR, the C release patterns showed significant ($F_{12, 40} = 3.67$, $P < 0.001$) differences among the decomposed litters during the course of the incubation. The C released from mixed cocoa-shade was the slowest and the cocoa litter released the greatest amount of C during the decomposition process (**Figure 3**WR: (C)). However, the % C released from the cocoa

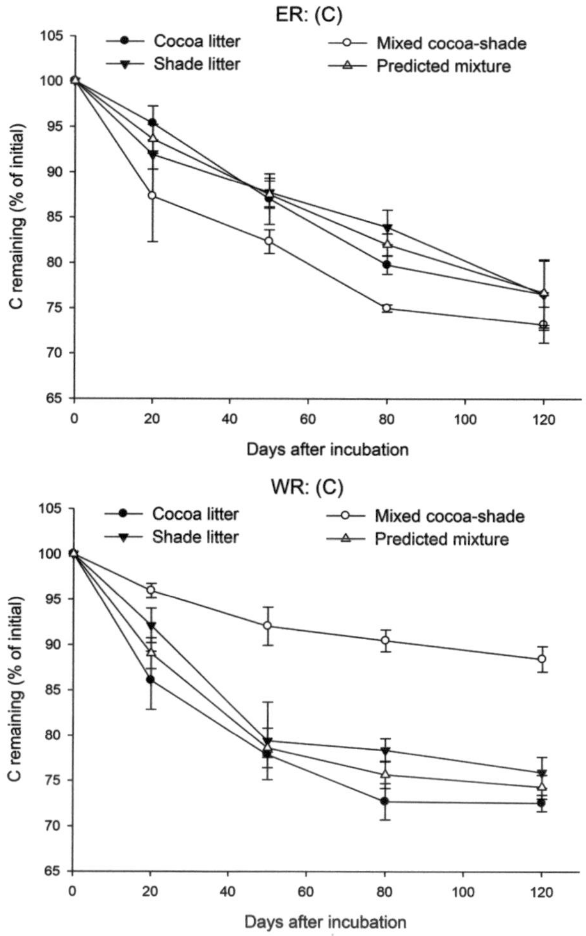

Figure 3.
Carbon released patterns (% of initial C) during a 120-day decomposition of leaf litters from cocoa ecosystems at various stages.

litters were not significantly ($P > 0.05$) different from those of shade and predicted mixture. The % carbon emitted after the 120 days of incubation varied from 11.5% in the mixed cocoa-shade litter to 27.4% in cocoa leaf litters. The implication of the slower C releasing rate of mixed litter than the expected is that, nutrient cycling through decomposition under shaded cocoa system would lead to nutrient limitation and, most likely, could cause nutrient deficiency for growing crops.

3.3 Carbon emission patterns during leaf decomposition

Decomposition processes in agro-ecosystems have been implicated as enriching the atmosphere with carbon dioxide (CO_2). During leaf litter decomposition, the decomposing litter is accompanied with losses of carbon. Although the amount of litter decomposed would be expected to be proportional to the C loss, considerable variations do occur due to different biochemical composition of different plant species [83]. There have not been many studies to monitor the fate of the C loss through the decomposed litter. This requires a method that will capture the C released as litter undergoes decomposition. If C loss data from such a method tends to be comparable to those from the litterbag technique, then the problem of having to retrieve decomposing litter is overcome.

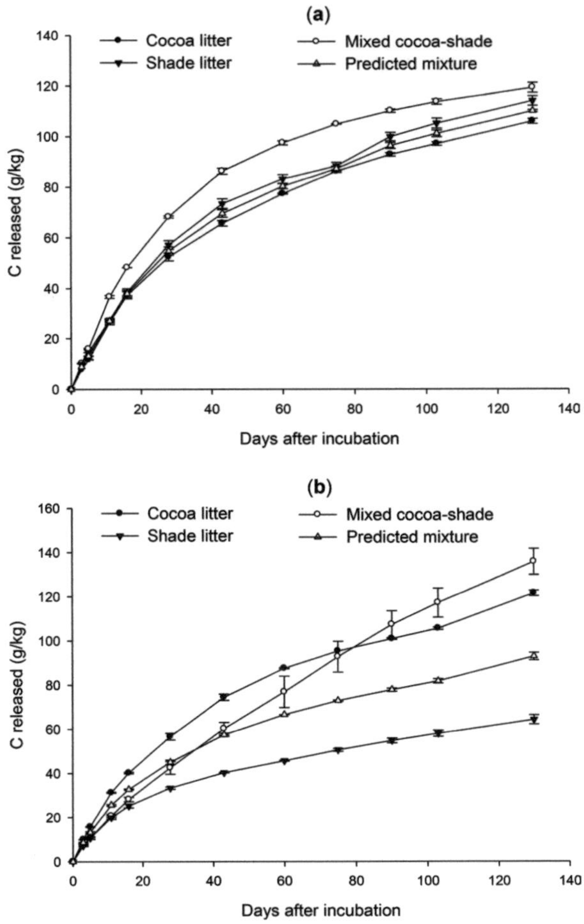

Figure 4.
Carbon mineralization patterns of leaf litters in cocoa ecosystems: (a) Eastern region and (b) Western region. Bars indicate standard error (n = 3).

As a response to the above concern, **Figure 4a** and **b** present the patterns and cumulative amounts of CO_2-C emissions measured during the decomposition of leaf litters collected under cocoa ecosystems of ER and WR of Ghana. The patterns of C emitted from the decomposing leaf litter were somewhat similar. Overall, C emissions increased rapidly during the first 16 days followed by a relatively slow rate as time progressed (**Figure 4a** and **b**). Previous works have also shown double-phase patterns for CO_2-C emissions during soil organic carbon mineralization [82, 85–88]. Even though the C emission patterns were similar, the litter treatments differed significantly ($P < 0.05$) in the amounts of C released at all times during the incubation.

Among the litter treatments from the ER, the mixed cocoa-shade consistently released significantly more C than the other treatments which did not differ in their C emissions (**Figure 4a**). Thus, whilst there was no difference between the pure cocoa and shade treatments, the C emission of their mixture was higher than expected and could not be predicted from the separate litter treatments as indicated by the significantly ($P < 0.05$) lower C emission pattern of the predicted mixture. At 130 days of incubation, the amounts of C emitted from the litter treatments ranged from 106.0 to 119.2 g/kg for cocoa and mixed cocoa-shade treatments, respectively (**Figure 4a**). The higher C mineralization in mixed cocoa-shade appeared to be the effect of P which is the only nutrient element that differed between the mixed cocoa-shade and the other samples from ER (**Table 2**).

Leaf treatments from the WR exhibited similar C release patterns as earlier stated; each showing a double phase comprising of an initial rapid rate and a subsequent decreasing rate as the incubation advanced (**Figure 4b**). However, in the case of the litter treatments from WR, the cocoa leaf treatment emitted significantly ($P < 0.05$) more C than the other treatments until 43 days of incubation when there was no longer significant difference from the C emissions of mixed cocoa-shade treatment (**Figure 4b**). The lowest C emission pattern was observed from the treatment with shade leaf litter. The C emission patterns of the mixed cocoa-shade and predicted mixture treatments exhibited fluctuations such that higher C emission was estimated by the predicted mixture from 5 to 16 days. The C emitted from both did not differ between 28 and 75 days, and thereafter, the C emissions of the mixed cocoa-shade treatment exceeded the predicted emissions towards the end of the observation (**Figure 4b**).

At 130 days, the range of C emitted from the treatment with leaf litters from WR was 64.3 to 135.8 g/kg for the shade and mixed cocoa-shade treated litters. Previous works on soil carbon mineralization reported that the initial biochemical composition plays a major role in driving the process and C/N ratio of decomposing organic material is said to be a good indicator of its mineralization potential [89]. However, the findings herein are unable to confirm or deny the importance of C/N ratio since the litters did not differ significantly and were all less than the critical C/N ratio of 25 below which mineralization would be expected [62, 63].

3.3.1 Potential mineralizable C for 130 days incubation

The potential mineralizable C pools from the leaf litters at the end of 130 days of incubation were estimated by fitting the C emission patterns (**Figure 4**) to a single exponential rise-to-maximum (growth) model (Eq. (6)). The same model has been used previously on similar data by others [82, 87, 90]. The data on emitted C conformed well to the model (R^2 = 98.0–99.9) and the estimated parameters provided estimates of potential mineralizable C and the mineralization rate constants of the litter treatments as presented in **Table 3**.

The potential mineralizable C (C_o) estimated for the litters from ER during the 130-day incubation had a narrow range of 107.7–118.5 g/kg for cocoa and mixed cocoa-shade treatments, respectively (**Table 3**). The estimated potential mineralizable C represents approximately 26.4–29.6% of the oxidizable C in leaf litters of cocoa and mixed cocoa-shade, respectively. This indicates that the leaf litters had close potential for releasing similar amounts of C. However, these estimated mineralizable carbon values differed significantly ($F_{3,4} = 11.50$, $P = 0.020$) according to the litter treatments. The differences in amount of estimated mineralizable C appear to reflect the quality of the oxidizable carbon source. Indeed, a correlation analysis indicated the presence of significant relationships between the estimated mineralizable C of leaf litters from ER and some initial chemical properties of the litter treatment as follows: N ($r = 0.713$, $P = 0.047$), P ($r = -0.784$, $P = 0.021$), and C/N ($r = -0.883$, $P = 0.004$). However, other researchers found high mineralizable C from decomposing *Mucuna* litter and attributed this to its low lignin content rather than its C/N ratio [85, 87]. This means that the amount of potential mineralizable C from decomposing organic material is partly controlled by its biochemical quality.

There were considerable variations in the estimated mineralizable C among the leaf litter obtained from the WR (**Table 3**). The results indicated a wider range of mineralizable C pools of 61.10–209.70 g/kg respectively for the shade and mixed cocoa-shade litter treatments. The estimated potential mineralizable C range for the WR litters represented a potential C loss range of 13.4–49.6% of the initial oxidizable C content of the litters within 130 days of incubation. The present estimate for C loss relative to the period is much higher when compared with Saffigna et al. [91] who reported a decline in mineralizable C by 29% when sorghum residues were removed for 6 years from a hitherto amended soil. However the lower C loss associated with sorghum residue partly reflects its lower oxidizable carbon content relative to cocoa litter. The cocoa litter contained approximately twice as much mineralizable C as contained in the shade litter, but the mixed cocoa-shade contained more than twice as much potential mineralizable C as expected by the predicted mixture capacity (**Table 3**).

There were no significant correlations between the estimated mineralizable C pools from the WR leaf litter treatments and the biochemical composition of the decomposing litter although there were indications of moderate relationship with each one of the following: C ($r = -0.672$, $P = 0.068$), N ($r = -0.564$, $P = 0.145$), and P ($r = 0.530$, $P = 0.176$). It thus confirms that the amount of potential mineralizable C estimated from the litter incubation partly reflected the initial chemical composition and the amount of oxidizable C in the decomposing litter.

3.4 Comparison of C from leaf weight loss and CO_2-C evolution methods

Although leaf decomposition is generally measured by weight loss, it is also measured by carbon dioxide release in numerous studies [24, 25]. These methods have several sources of variations as mentioned in the introductory section that potentially could confound the results and cause deviations from litter decomposition under natural vegetation types. Comparison of methods is an option through which interference from the methods with the results can be isolated.

Table 5 presents the measured amounts of C released from cocoa systems in ER and WR during leaf litter decomposition averaged over the incubation period (120 days) by the weight loss and carbon dioxide evolution methods. Under the cocoa systems in the Eastern region, the C released measured by the CO_2 evolution method was significantly higher than measured by the weight loss method in cocoa leaf ($F_{1,3} = 38.1$, $P = 0.009$), mixed litter ($F_{1,3} = 23.5$, $P = 0.017$) and predicted mixed litter ($F_{1,3} = 18.4$, $P = 0.023$) decomposition but their difference was not significant ($F_{1,3} = 0.07$, $P = 0.810$) with respect to shade tree leaf decomposition (**Table 5**). These higher amounts of C from the released CO_2 measurements are usually unexpected under natural unconfined environments since the release of CO_2

Region	Litter			
	Cocoa leaf litter	Shade leaf litter	Mixed leaf litter	Predicted mixed litter
Eastern region				
Loss litter C	167.7b	329.0a	230.8b	248.4b
Evolved CO_2-C	304.0a	326.4a	371.1a	315.2a
Western region				
Loss litter C	371.3a	153.8b	324.7a	262.5a
Evolved CO_2-C	338.3a	185.1a	326.9a	261.8a

[1]*Different letters within same region and column indicate significant difference at* P < 0.05 *using Tukey's method.*

Table 5.
Cumulative C released (g/kg) over 120 days during litter decomposition under cocoa systems in Eastern and Western regions as measured by the weight loss and CO_2-C evolution methods.[1]

during litter decomposition is but one of several ways including fragmentation and organic matter leaching that contributes to the weight loss of buried litter [24, 25].

However, the confinement of the current experiment as described earlier meant that weight losses through fragmentation and leaching of organic matter were disallowed. Therefore, the litter weight loss solely depended on the release amounts of CO_2-C during the period. Thus, the expected weight loss from decomposing litters must at most, be equivalent to the amounts of CO_2 released in the absence of leaching of other organic material. The many steps such as initial total C determination of the decomposing litter, weighing litter remains that have not been dried well or have attached soil particles to determine weight loss and the use of larger time intervals are all sources of variations associated with the determination of C release by the weight loss method. These steps have the potential of being over estimated and might explain the lower amounts C losses by the weight loss method when compared to the CO_2-C evolution method. On the other hand, the short time intervals of the CO_2-C evolution methods with regard to the frequent replacement of the adsorbent creates room for atmospheric CO_2-C to interfere with the decomposition and consequently leads to over estimation of release C from the litter decomposition alone.

Apparently, the above sources of deviations associated with the two methods were minimal under the litter treatments of WR cocoa systems. Hence, the expectation of equality between litter weight loss C and CO_2-C released was confirmed in all but the shade tree leaf litter decomposition. Differences between released C estimated from litter weight loss and CO_2-C evolution measurement methods were not statistically significant under decomposing cocoa litter ($F_{1,3}$ = 3.9, P = 0.143), mixed litter ($F_{1,3}$ = 0.008, P = 0.933) and predicted litter ($F_{1,3}$ = 0.008, P = 0.934) in the WR (**Table 5**). In contrast, the shade tree leaf litter decomposition from WR released significantly ($F_{1,3}$ = 11.5, P = 0.043) different amounts of C, where the measured C by CO_2-C evolved was higher than that in loss litter (**Table 5**).

Although the measured differences varied between regions, regression analysis of pooled data from the two regions indicated the existence of a strong relationship between weight loss of litter C and the CO_2-C evolution during litter decomposition (**Figure 5**). The line of best fit to the scatter suggests that CO_2-C emission is proportional to litter decomposition in the cocoa ecosystems. Quantitatively, litter decomposition accounts for 70.9% of the variations in measured CO_2-C emissions from the cocoa ecosystems (**Figure 5**). Other researchers have also found strong agreements between the two methods with respect to measuring the C released during decomposition [27, 31, 32].

3.5 Mitigation of CO_2 emissions

The quantity of carbon released from the decomposition of dead materials into the atmosphere contributes significantly to the global carbon budget. It is estimated that about 70% total annual carbon flux (this is equivalent to 68 Pg C/y) derives from the decomposition of plant materials [92]. Forests are recognized as an important component for climate mitigation and adaptation. Conceivably, promoting agroforestry practices such as cocoa ecosystems in the tropics on cleared lands would mitigate the atmospheric CO_2 loads through photosynthesis and C storage in their tissues. The amount of C stored is proportional to the biomass of the tree components and consequently the amount of CO_2 removed from the atmosphere.

In comparing the C stored in cocoa systems with annual crops, many studies have reported higher C storage in the cocoa systems [22, 93, 94]. Lavelle and Pashanasi [95] noted that forest ecosystems and pastures contain more biomass C than cropland. On a vertisol in Ethiopia, Lulu and Insam [96] observed positive

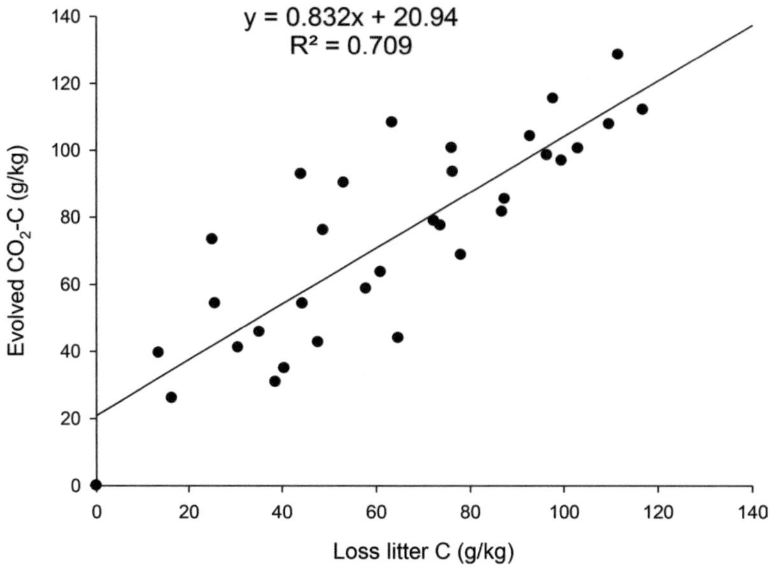

Figure 5.
Relationship between measurements of C released during litter decomposition by the litter weight loss and the CO_2-C evolution methods.

effects of agroforestry practice with Sesbania on soil organic carbon (SOC) pool. Dowuona et al. [97] reported a 25.6 g/kg SOC on a ferric acrisol under *Leucaena leucocephala* woodlot in Ghana compared to the 15.6 g/kg SOC for its *Chromolaena odorata* native fallow adjacent soil.

The recognition of the potential of sequestering carbon in plantations has attracted the attention of many researchers on C sequestration projects. These researchers have predicted a potential market for C in developing nations as a result of the investments from companies and governments wishing to offset their emissions of greenhouse gases as directed by the Kyoto Protocol's Clean Development Mechanism [98, 99].

Whether the soil acts as a source or sink of carbon gases depends greatly on the type and intensity of activities of human management on the land. Soil management practices have been documented to have tremendous effects on soil organic matter (SOM) storage. In a study from adjacent forested and cultivated soils in eight agro-ecosystems from the Ethiopian highlands and Nigerian lowlands, SOM content was two to four times higher in the forested than in the cultivated soils [100]. In an 11-year experiment to assess the potential of different cropping systems to sequester C in the soils, Bostick et al. [101] noted significant reductions of soil organic carbon from a continuous fallow of 0.53% C to 0.46, 0.37, 0.35 and 0.33% C for sorghum-fallow, continuous cotton, continuous sorghum and cotton-maize-sorghum rotations, respectively, in Burkina Faso. Haynes and Francis [102] have reported high amounts of C under pasture relative to cultivated soils. Pichot et al. [103] observed that average soil C increased between 116 and 377 kg/ha/y in a 10-year study in Burkina Faso when soils were amended with low and high levels of inorganic and organic fertilizer, respectively.

4. Conclusions

Litter decomposition helps to replenish soil nutrient pools. Therefore, plant litter decomposition plays a key role in biogeochemical nutrient cycling, the rate of which determines the productivity of natural and in part agro ecosystems. The findings of this study have contributed to our understanding of litter decomposition and C dynamics in cocoa ecosystems of Ghana.

Trends of leaf litter decomposition and C mineralization indicated that mixed cocoa-shade litter treatments decomposed faster than the cocoa leaf litter alone; this suggests that litter mixing has a positive interaction effect in cocoa ecosystems. The management implication of this finding is that if the release of nutrients into the soil is a consequence of litter decomposition, then the mixed litter systems as in shaded cocoa ecosystems would be more effective in releasing plant nutrients than the single tree species litter systems.

Author details

Askia M. Mohammed[1*], James S. Robinson[2], David J. Midmore[2] and Anne Verhoef[2]

1 CSIR-Savanna Agricultural Research Institute, Tamale, Ghana

2 University of Reading, UK

*Address all correspondence to: mamusah@yahoo.com

References

[1] Bandaranayake W, Qian Y, Parton W, Ojima D, Follett R. Estimation of soil organic carbon changes in turfgrass systems using the CENTURY model. Agronomy Journal. 2003;**95**:558-563

[2] Howard P, Howard D. Microbial decomposition of tree and shrub leaf litter. 1. Weight loss and chemical composition of decomposing litter. Oikos. 1974;**25**:341-352

[3] Vitousek PM, Aber JD, Howarth RW, Likens GE, Matson PA, Schindler DW, et al. Human alteration of the global nitrogen cycle: Sources and consequences. Ecological Applications. 1997;**7**:737-750

[4] Houghton RA. The annual net flux of carbon to the atmosphere from changes in land use 1850-1990*. Tellus B. 1999;**51**:298-313

[5] Walker BH, Steffen WL. The nature of global change. In: Walker B, Steffen W, Canadell J, Ingram J, editors. The Terrestrial Biosphere and Global Change. Cambridge, UK: Cambridge University Press; 1999. p. 18

[6] Jaramillo VJ, Kauffman JB, Renteria-Rodriguez L, Cummings DL, Ellingson LJ. Biomass, carbon, and nitrogen pools in Mexican tropical dry forest landscapes. Ecosystems [Internet]. 2003;**6**(7):609-629. Available from: http://www.springerlink.com/openurl.asp?genre=article&id=doi:10.1007/s10021-002-0195-4

[7] Houghton RA, Hackler JL. Carbon flux to the atmosphere from land-use changes. Trends: A Compendium of Data on Global Change. 2001;**51**:1-18

[8] Brown S. Measuring carbon in forests: Current status and future challenges. Environmental Pollution. 2002;**116**(3):363-372. Available from: http://www.ncbi.nlm.nih.gov/pubmed/11822714

[9] Sharrow S, Ismail S. Carbon and nitrogen storage in agroforests, tree plantations, and pastures in western Oregon, USA. Agroforestry Systems. 2004;**60**:123-130

[10] Trinsoutrot I, Recous S, Bentz B, Lineres M, Cheneby D, Nicolardot B. Biochemical quality of crop residues and carbon and nitrogen mineralization kinetics under nonlimiting nitrogen conditions. Soil Science Society of America Journal. 2000;**64**:918-926

[11] Kumar K, Goh K. Nitrogen release from crop residues and organic amendments as affected by biochemical composition. Communications in Soil Science and Plant Analysis. 2003;**34**:2441-2460

[12] Palm C, Sanchez P. Nitrogen release from the leaves of some tropical legumes as affected by their lignin and polyphenolic contents. Soil Biology and Biochemistry. 1991;**23**:83-88

[13] Tian G, Brussaard L, Kang B. An index for assessing the quality of plant residues and evaluating their effects on soil and crop in the (sub-) humid tropics. Applied Soil Ecology. 1995;**2**:25-32

[14] Vanlauwe B, Diels J, Sanginga N, Merckx R. Residue quality and decompostion: An unsteady relationship? Driven by Nature: Plant Litter Quality and Decomposition. 1996:157-166

[15] Vossbrinck CR, Coleman DC, Wooley TA. Abiotic and biotic factors in litter decomposition in semi-arid grassland. Ecology. 1979;**60**:265-271

[16] Anglaaere LCN. Improving the Sustainability of cocoa Farms in Ghana

through Utilization of Native Forest Trees in Agroforestry Systems. Bangor, UK: University of Wales; 2005

[17] Ling A. Litter Production and nutrient Cycling in a Mature cocoa Plantation on Inland Soils of Peninsular Malaysia, Cocoa and Coconuts: Progress and Outlook. Kuala Lumpur: Inc. Society of Planters; 1986. pp. 451-466

[18] Aranguren J, Escalante G, Herrera R. Nitrogen cycle of tropical perennial crops under shade trees. Plant and Soil. 1982;**67**:247-258

[19] Heuveldop J, Fassbender HW, Alpizar L, Enriquez G, Fölster H. Modelling agroforestry systems of cacao (*Theobroma cacao*) with laurel (*Cordia alliodora*) and poro (*Erythrina poeppigiana*) in Costa Rica II. Cacao and wood production, litter production and decomposition. Agroforestry Systems. 1988;**6**:37-48

[20] de Oliveira Leite J, Valle RR. Nutrient cycling in the cacao ecosystem: Rain and throughfall as nutrient sources for the soil and the cacao tree. Agriculture, Ecosystems and Environment. 1990;**32**:143-154

[21] Wessel M. Shade and nutrition of cocoa. In: Hartemink AE. (2005), Nutrient Stocks, Nutrient Cycling, and Changes in Cocoa Ecosystems: A Review. Advances in Agronomy. 1985;**86**:227-253

[22] Hartemink AE. Nutrient stocks, nutrient cycling, and soil changes in cocoa ecosystems: A review. Advances. 2005;**86**:227-253

[23] Frankland JC. Succession of fungi on decaying petioles of *Pteridium aquilinum*. Journal of Ecology. 1966;**54**:41-63

[24] Okeke AI, Omaliko C. Leaf litter decomposition and carbon dioxide evolution of some agroforestry fallow species in southern Nigeria. Forest Ecology and Management. 1992;**50**:103-116

[25] Quemada M, Cabrera M. Carbon and nitrogen mineralized from leaves and stems of four cover crops. Soil Science Society of America Journal. 1995;**59**:471-477

[26] Bernhard-Reversat F. Changes in relationships between initial litter quality and CO_2 release during early laboratory decomposition of tropical leaf litters. European Journal of Soil Biology. 1998;**34**:117-122

[27] Ofori-Frimpong K, Rowell DL. The decomposition of cocoa leaves and their effect on phosphorus dynamics in tropical soil. European Journal of Soil Science. 1999;**50**(1):165-172. Available from: http://doi.wiley.com/10.1046/j.1365-2389.1999.00208.x

[28] Hoorens B, Aerts R, Stroetenga M. Litter quality and interactive effects in litter mixtures: More negative interactions under elevated CO_2? Journal of Ecology. 2002;**90**:1009-1016

[29] Fening J, Adjei-Gyapong T, Yeboah E, Ampontuah E, Quansah G, Danso S. Soil fertility status and potential organic inputs for improving small holder crop production in the interior savanna zone of Ghana. Journal of Sustainable Agriculture. 2005;**25**:69-92

[30] Marstorp H. Influence of soluble carbohydrates, free amino acids, and protein content on the decomposition of *Lolium multiflorum* shoots. Biology and Fertility of Soils. 1996;**21**:257-263

[31] Briones M, Ineson P. Decomposition of eucalyptus leaves in litter mixtures. Soil Biology and Biochemistry. 1996;**28**:1381-1388

[32] Murungu FS, Chiduza C, Muchaonyerwa P, Mnkeni PNS. Decomposition, nitrogen and

phosphorus mineralization from winter-grown cover crop residues and suitability for a smallholder farming system in South Africa. Nutrient Cycling in Agroecosystems. 2011;**89**(1): 115-123. Available from: http://www.springerlink.com/index/10.1007/s10705-010-9381-5

[33] Meentemeyer V. Macroclimate and lignin control of litter decomposition rates. Ecology. 1978;**59**:465-472

[34] Swift MJ, Heal OW, Anderson JM. Decomposition in Terrestrial Ecosystems. Berkeley, Los Angeles: University of California Press; 1979

[35] Aber JD, Melillo JM. Litter decomposition: Measuring relative contributions of organic matter and nitrogen to forest soils. Canadian Journal of Botany. 1980;**58**:416-421

[36] Aerts R. Climate, leaf litter chemistry and leaf litter decomposition in terrestrial ecosystems: A triangular relationship. Oikos. 1997;**79**:439-449

[37] Pereira AP, Graca MAS, Molles M. Leaf litter decomposition in relation to litter physio-chemical properties, fungal biomass, arthropod colonization, and geographical origin of plant species. Pedobiologia (Jena). 1998;**42**:316-327

[38] Lill JT, Marquis RJ. The effects of leaf quality on herbivore performance and attack from natural enemies. Oecologia. 2001;**126**:418-428

[39] Thomas WA. Decomposition of loblolly pine needles with and without addition of dogwood leaves. Ecology. 1968;**49**:568-571

[40] Staaf H. Influence of chemical composition, addition of raspberry leaves, and nitrogen supply on decomposition rate and dynamics of nitrogen and phosphorus in beech leaf litter. Oikos. 1980;**35**:55-62

[41] Seastedt T. The role of microarthropods in decomposition and mineralization processes. Annual Review of Entomology. 1984;**29**:25-46

[42] Gartner TB, Cardon ZG. Decomposition dynamics in mixed species leaf litter. Oikos. 2004;**104**:230-246

[43] Hansen RA, Coleman DC. Litter complexity and composition are determinants of the diversity and species composition of oribatid mites (Acari: Oribatida) in litterbags. Applied Soil Ecology. 1998;**9**:17-23

[44] Biederlack L, Rivers J. Comprehensive Food Security and Vulnerability Analysis (CFSVA). Ghana: United Nations World Food Programme; 2009

[45] Zarcinas B, Cartwright B, Spouncer L. Nitric acid digestion and multi-element analysis of plant material by inductively coupled plasma spectrometry. Communications in Soil Science and Plant Analysis. 1987;**18**:131-146

[46] Anderson JM, Ingram JSI. Tropical Soil Biology and Fertility: A Handbook of Methods. Oxon: CAB International; 1993

[47] Rowell D. Soil Science: Methods and Applications. Singapore: Longman Singapore Publishers (Pte) Ltd; 1994

[48] Liu P, Huang J, Sun OJ, Han X. Litter decomposition and nutrient release as affected by soil nitrogen availability and litter quality in a semiarid grassland ecosystem. Oecologia. 2010;**162**:771-780

[49] Tan Y, Chen JY, An L, Huang J, Wang L, Chen S. Mass loss and nutrient dynamics during litter decomposition under three mixing treatments in a typical steppe in Inner Mongolia. Plant and Soil. 2013;**366**:107-118

[50] Olson JS. Energy storage and the balance of producers and decomposers in ecological systems. Ecology. 1963;**44**:322-331

[51] Oglesby KA, Fownes JH. Effects of chemical composition on nitrogen mineralization from green manures of seven tropical leguminous trees. Plant and Soil. 1992;**143**:127-132

[52] Kemp P, Reynolds J, Virginia R, Whitford W. Decomposition of leaf and root litter of Chihuahuan desert shrubs: Effects of three years of summer drought. Journal of Arid Environments. 2003;**53**:21-39

[53] Dossa E, Khouma M, Diedhiou I, Sene M, Kizito F, Badiane A, et al. Carbon, nitrogen and phosphorus mineralization potential of semiarid Sahelian soils amended with native shrub residues. Geoderma. 2009;**148**:251-260

[54] Honeycutt C, Potaro L, Avila K, Halteman W. Residue quality, loading rate and soil temperature relations with hairy vetch (Vicia villosa Roth) residue carbon, nitrogen and phosphorus mineralization. Biological Agriculture and Horticulture. 1993;**9**:181-199

[55] Woomer P, Martin A, Albrecht A, Resck D, Scharpenseel H, Swift M. The Importance and Management of Soil Organic Matter in the Tropics. Hoboken, New Jersey: John Wiley & Sons; 1994

[56] Kuo S, Sainju U, Jellum E. Winter cover cropping influence on nitrogen in soil. Soil Science Society of America Journal. 1997;**61**:1392-1399

[57] Brown S. Estimating Biomass and Biomass Change of Tropical Forests: A primer (FAO Forestry Paper–134). Rome: Food and Agriculture Organization; 1997. Available from: http://www.fao.org/docrep/W4095E/w4095e00.htm

[58] Montagnini F, Porras C. Evaluating the role of plantations as carbon sinks: An example of an integrative approach from the humid tropics. Environmental Management. 1998;**22**:459-470

[59] Snowdon P. Synthesis of Allometrics, Review of Root Biomass and Design of Future Woody Biomass Sampling Strategies. Australia: Australian Greenhouse Office; 2000

[60] IPCC. Good Practice Guidelines for National Greenhouse Gas Inventories. Switzerland: IPCC; 2006

[61] Egbe N, Olatoye S, Obatolu C. Impact of rate and types of fertilizers on productivity and nutrient cycling in tree crop plantation ecosystem. In: Unpublished Paper Submitted for MAB Workshop. 1989

[62] Nair PR. An Introduction to Agroforestry. Dordrecht, The Nertherlands: Kluwer Academic Publisher; 1993

[63] Sainju UM, Whitehead WF, Singh BP. Biculture legume–cereal cover crops for enhanced biomass yield and carbon and nitrogen. Agronomy Journal. 2005;**97**:1403-1412

[64] Girisha G, Condron L, Clinton P, Davis M. Decomposition and nutrient dynamics of green and freshly fallen radiata pine (*Pinus radiata*) needles. Forest Ecology and Management. 2003;**179**:169-181

[65] Berg B, Staaf H. Leaching, accumulation and release of nitrogen in decomposing forest litter. Ecological Bulletins. 1981;**33**:163-178

[66] Blair G, Boland O. The release of phosphorus from plant material added to soil. Soil Research. 1978;**16**:101-111

[67] Mafongoya PL, Barak P, Reed JD. Carbon, nitrogen and phosphorus mineralization of tree leaves and

manure. Biology and Fertility of Soils. 2000;**30**:298-305

[68] Aikpokpodion PE. Nutrients dynamics in cocoa soils, leaf, and beans in Ondo State, Nigeria. The Journal of Agricultural Science. 2010;**1**(1):1-9

[69] Dawoe E. Conversion of Natural Forest to Cocoa Agroforest in Lowland Humid Ghana: Impact on Plant Biomass Production, Organic Carbon and Nutrient Dynamics. Kwame Nkrumah University of Science and Technology; 2009

[70] Mafongoya P, Nair P, Dzowela B. Mineralization of nitrogen from decomposing leaves of multipurpose trees as affected by their chemical composition. Biology and Fertility of Soils. 1998;**27**:143-148

[71] Dawoe EK, Isaac ME, Quashie-Sam J. Litterfall and litter nutrient dynamics under cocoa ecosystems in lowland humid Ghana. Plant Soil. 2009;**330** (1-2):55-64. Available from: http://www.springerlink.com/index/10.1007/s11104-009-0173-0

[72] Baggie I, Rowell D, Robinson J, Warren G. Decomposition and phosphorus release from organic residues as affected by residue quality and added inorganic phosphorus. Agroforestry Systems. 2005;**63**:125-131

[73] Scherer-Lorenzen M, Luis Bonilla J, Potvin C. Tree species richness affects litter production and decomposition rates in a tropical biodiversity experiment. Oikos. 2007;**116**:2108-2124

[74] Mohammed AM, Nartey EK, Naab JBB, Adiku SGK. A simple model for predicting plant residue decomposition based upon their C/N ratio and soil water content. African Journal of Agricultural Research. 2013;**8**(19):2153-2159

[75] Anim-Kwapong GJGJ, Osei-Bonsu K. Potential of natural and improved fallow using indigenous trees to facilitate cacao replanting in Ghana. Agroforestry Systems. 2009;**76**(3):533-542. Available from: http://www.springerlink.com/index/10.1007/s10457-008-9196-4

[76] Owusu-Sekyere EE, Nakashima K, Wakatsuki T. Extending cocoa agroforestry into Sawah ecosystem in Ghanaian inland valleys. Ghana Journal of Agricultural Science. 2010;**43**:37-44

[77] Scherer-Lorenzen M, Elend A, Nöllert S, Schulze E-D. Plant Invasions in Germany–General Aspects and Impact Of Nitrogen Deposition, Invasive Species in a Changing World. Covelo: Island Press; 2000. pp. 351-368

[78] Uchida R. Essential nutrients for plant growth: Nutrient functions and deficiency symptoms. Plant Nutrient Management in Hawaii's Soils. 2000:31-55

[79] Murwira H, Kirchmann H, Swift M. The effect of moisture on the decomposition rate of cattle manure. Plant and Soil. 1990;**122**:197-199

[80] Saviozzi A, Levi-Minzi R, Riffaldi R. Mineralization parameters from organic materials added to soil as a function of their chemical composition. Bioresource Technology. 1993;**45**:131-135

[81] Pascual J, Hernandez T, García C, Ayuso M. Carbon mineralization in an arid soil amended with organic wastes of varying degrees of stability. Communications in Soil Science and Plant Analysis. 1998;**29**:835-846

[82] Nourbakhsh F, Sheikh-Hosseini AR. A kinetic approach to evaluate salinity effects on carbon mineralization in a plant residue-amended soil. Journal of Zhejiang University. Science. B. 2006;**7**:788-793

[83] McTiernan KB, Ineson P, Coward PA. Respiration and nutrient release

from tree leaf litter mixtures. Oikos. 1997:527-538

[84] Vitousek PM, Turner DR, Parton WJ, Sanford RL. Litter decomposition on the Mauna Loa environmental matrix, Hawai'i: Patterns, mechanisms, and models. Ecology. 1994;**75**:418-429

[85] Nourbakhsh F. Fate of carbon and nitrogen from plant residue decomposition in a calcareous soil. Plant, Soil and Environment. 2006;**52**:137-146

[86] Bonde TA, Rosswall T. Seasonal variation of potentially mineralizable nitrogen in four cropping systems. Soil Science Society of America Journal. 1987;**51**:1508-1514

[87] Ajwa H, Tabatabai M. Decomposition of different organic materials in soils. Biology and Fertility of Soils. 1994;**18**:175-182

[88] Riffaldi R, Saviozzi A, Levi-Minzi R. Carbon mineralization kinetics as influenced by soil properties. Biology and Fertility of Soils. 1996;**22**:293-298

[89] Cheruiyot E, Mwonga S, Mumera L, Macharia J, Tabu I, Ngugi J. Rapid decay of dolichos [*Lablab purpureus* (L.) Sweet] residue leads to loss of nitrogen benefit to succeeding maize (*Zea mays* L.). Animal Production Science. 2007;**47**:1000-1007

[90] Martinez C, Tabatabai M. Decomposition of biotechnology by-products in soils. Journal of Environmental Quality. 1997;**26**:625-632

[91] Saffigna P, Powlson D, Brookes P, Thomas G. Influence of sorghum residues and tillage on soil organic matter and soil microbial biomass in an Australian vertisol. Soil Biology and Biochemistry. 1989;**21**:759-765

[92] Raich JW, Schlesinger WH. The global carbon dioxide flux in soil respiration and its relationship to vegetation and climate. Tellus B. 1992;**44**:81-99. Available from: http://www.blackwell-synergy.com/links/doi/10.1034%2Fj.1600-0889.1992.t01-1-00001.x

[93] Kotto-same J, Woomer PL, Appolinaire M, Louis Z. Carbon dynamics in slash-and-bum agriculture and land use alternatives of the humid forest zone in Cameroon. Science (80-). 1997;**809**:97

[94] Duguma B, Gockowski J, Bakala J. Smallholder cacao (*Theobroma cacao* Linn.) cultivation in agroforestry systems of West and Central Africa: Challenges and opportunities. Agroforestry Systems. 2001;**51**:177-188

[95] Lavelle P, Pashanasi B. Soil macrofauna and land management in Peruvian Amazonia. Pedobiologia (Jena). 1989;**33**:283-291

[96] Lulu B, Insam H. Effects of cropping systems and manuring on soil microbial and carbon status of a vertisol in Ethiopia. Bodenkultur. 2000;**51**:107-114

[97] Dowuona GN, Mermut AR, Adiku SGK, Nartey E, Tete-Mensah I. Improvements in the quality of soils under agroforestry practice in Ghana. Soil fertility management in West African land use systems. 1997;**34**:251-258

[98] Fearnside PM. Forests and global warming mitigation in Brazil: Opportunities in the Brazilian forest sector for responses to global warming. Biomass and Bioenergy. 1999;**16**:171-189

[99] Wise R, Cacho O. A bioeconomic analysis of carbon sequestration in farm forestry: A simulation study of *Gliricidia sepium*. Agroforestry Systems [Internet]. 2005;**64**(3):237-250. Available from: http://www.springerlink.com/index/10.1007/s10457-004-3938-8

[100] Spaccini R, Zena A, Igwe CA, Mbagwu JSC, Piccolo A. Carbohydrate in water-stable aggregates and particle size fractions of forested and cultivated soils in two contrasting tropical ecosystems. Biogeochemistry. 2001;**53**:1-22

[101] Bostick WM, Bado VB, Bationo A, Soler CT, Hoogenboom G, Jones JW. Soil carbon dynamics and crop residue yields of cropping systems in Northern Guinea Savanna of Burkina Faso. Soil and Tillage Research. 2006;**2257**:1-14

[102] Haynes RJ, Francis GS. Changes in microbial biomass C, soil carbohydrate composition and aggregate stability induced by growth of selected crop and forage species under field conditions. Journal of Soil Science. 1993;**44**:665-675

[103] Pichot J, Sedogo M, Poulain J, Arrivets J. Fertility evolution in a tropical ferruginous soil under the effect of organic manure and inorganic fertilizer applications [ploughed-in green manure, crop residues, straw; rotating sorghum-legume crops; aluminium toxicity allelopathy; Upper Volta]. Agronomia Tropical. 1981;**36**:122-133

Chapter 6

Greenhouse Gas Assessment and Strategies to Achieve CO_2 Sequestration in the Brazilian Palm Oil Life Cycle

Leidivan Almeida Frazão, Guilherme Silva Raucci, João Luis Nunes Carvalho, Marcelo Valadares Galdos, Cindy Silva Moreira, Carlos Eduardo Pellegrino Cerri and Carlos Clemente Cerri

Abstract

As the palm oil production is expanding in Brazilian Amazon region, this study aimed to determinate the greenhouse gas (GHG) emissions since the agricultural phase to transportation of crude palm oil (CPO) and then indicate strategies to achieve the CO_2 sequestration. The scope of this study comprised since the stage of oil palm seedlings production until the transportation of CPO. Inventory data for the year of 2009 included the agricultural production of fresh fruit bunches (FFB) and the extraction and transportation of CPO. The management of palm oil mill effluent (POME), use of fertilizers, fuels, pesticides, and electricity contributed to 66.5, 17.9, 15.1, 0.4, and 0.1% of the total emissions, respectively. Agricultural phase, CPO extraction, and transportation emitted 32,131, 79,590, and 1,104 t CO_2-eq, respectively. The carbon (C) footprint was 0.79 t CO_2-eq / t CPO, and the highest GHG emissions were associated to the management of POME. On the other hand, the use of all residues from the mill as fertilizer substitute can minimize the GHG emissions and increase soil C stocks. In addition, the methane (CH_4) from POME captured and used for steam or electricity is also a viable alternative to reduce the GHG emissions.

Keywords: carbon dioxide, crude palm oil, fresh fruit bunches, methane, nitrous oxide, palm oil mill effluent

1. Introduction

Oil palm (*Elaeis guineensis* Jacq.) is the most produced oil crop in the world due to its high productivity in relation to other oleaginous crops (e.g., soybean, sunflower, and rapeseed), and could meet growing global demand that is estimated to reach 240 million tons of palm oil by 2050 [1]. In Brazil, the cultivated area with oil palm is about 236,000 ha, including areas of agro-industries, small- and medium-sized farmers, family farmers and members of agrarian reform. It is estimated that

88% of Brazilian palm oil production was located at Pará State, while and Bahia and Roraima States account for 11 and 1% of production, respectively [2].

There is a great potential for development of palm oil chain in Brazil. The Brazilian government envisions a 35% expansion of oil palm production in the Amazon region due to its favorable soil and climate conditions and large amount of available lands. The Program for Sustainable Production of Palm Oil in Brazil has been promoting oil palm plantations but limiting the expansion only to degraded lands [3]. There was an encouragement for the cultivation of oil palm in the northern region due to its high productivity and potential for inclusion in the biodiesel agenda. Pará State continued to be targeted as the largest potential producer, and there was an estimated expansion of 330,000 ha until 2020 [4].

The environmental impacts of palm oil production can be accessed from a life cycle point of view, where the greenhouse gas (GHG) emissions can be accounted since from oil palm cultivation, crude palm oil (CPO) extraction, CPO transportation, and recycling or disposal of residues from the mill [5, 6]. According to Menichetti and Otto [7], the impacts depend on the consumption of conventional fuels, fertilizers, and the wastes generated. We hypothesized that the main source of GHG emissions in the palm oil production is related do CPO extraction and disposal of liquid waste from the mill.

Therefore, the aim of this study was to access the GHG emissions derived from the agricultural production, CPO extraction and transportation in an agro-industry farm located at Pará State, Amazon region, Brazil. Additionally, we aim to propose strategies for reducing GHG emissions and for promoting the CO_2 sequestration.

2. Materials and methods

2.1 Location of the study area

The study was carried out in 2009 in a commercial farm located in the municipality of Tailândia (48°46′W, 2°27′S), Pará State, Brazil (**Figure 1**). The company had a total area of 107,000 ha, where 60% is occupied with native vegetation, 36% with oil palm plantation, and 4% with infrastructure. The native vegetation is tropical rainforest. According to Köppen classification, the climate is Afi (tropical monsoonal). The rainfall for 2009 was 2705 mm year^{-1} and the mean temperature was 26.5°C. The temperature ranged from 22.6 to 33.4°C and the mean of annual relative humidity was higher throughout the year. The mean altitude of the region is 30 m, and the soil is well drained with medium clay content (18–29%), and classified as "Latossolo Amarelo distrófico típico" in the Brazilian System of Soil Classification [8] and as Oxisol (Xanthic Hapludox) in the USDA classification [9].

2.2 Palm oil system description

The palm oil supply chain consists of three subsystems (**Figure 2**): seedlings production, fresh fruit bunches (FFBs) production, and crude palm oil (CPO) extraction. Several processes are involved in the production of FFB, including planning, nursery establishment (seedlings cultivation), soil preparation, field establishment, field maintenance, harvesting and collection of FFB and replanting [10]. Considering all steps, the agricultural phase of oil palm production lasts for around 25–30 years [1].

The industrial phase consists of CPO extraction. The palm oil mill is located near to the oil palm plantations to facilitate the timely transportation and effective processing of FFB. Processing in the palm oil mill involves four major unit operations: sterilization, threshing and stripping of fruits, digestion, and oil extraction [11].

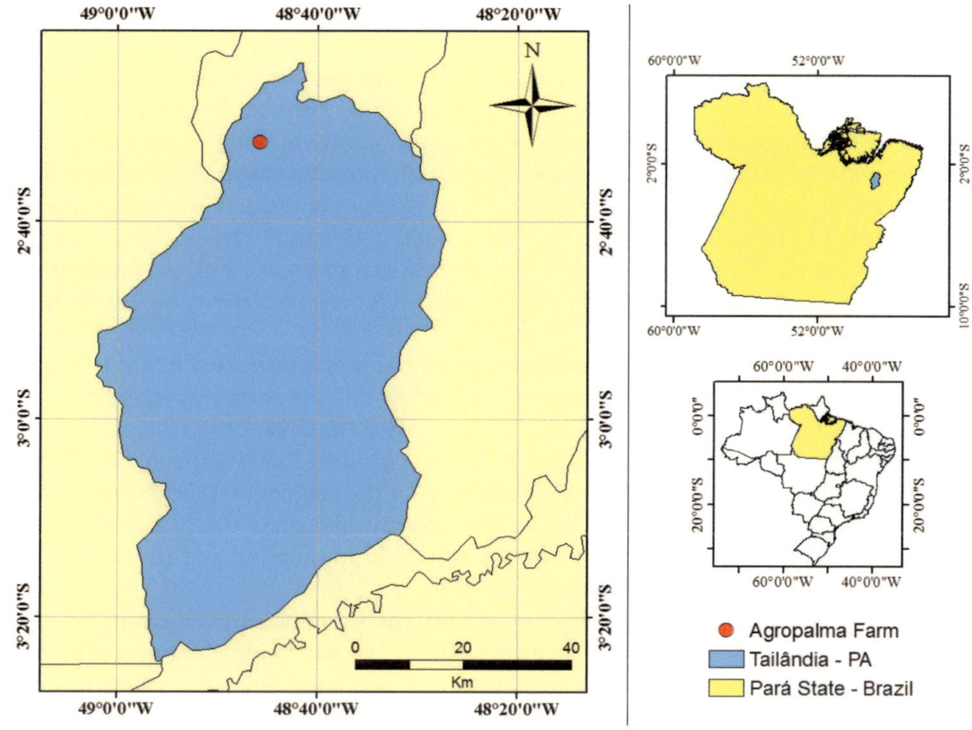

Figure 1.
Location map of the study area in the Brazilian Amazon region.

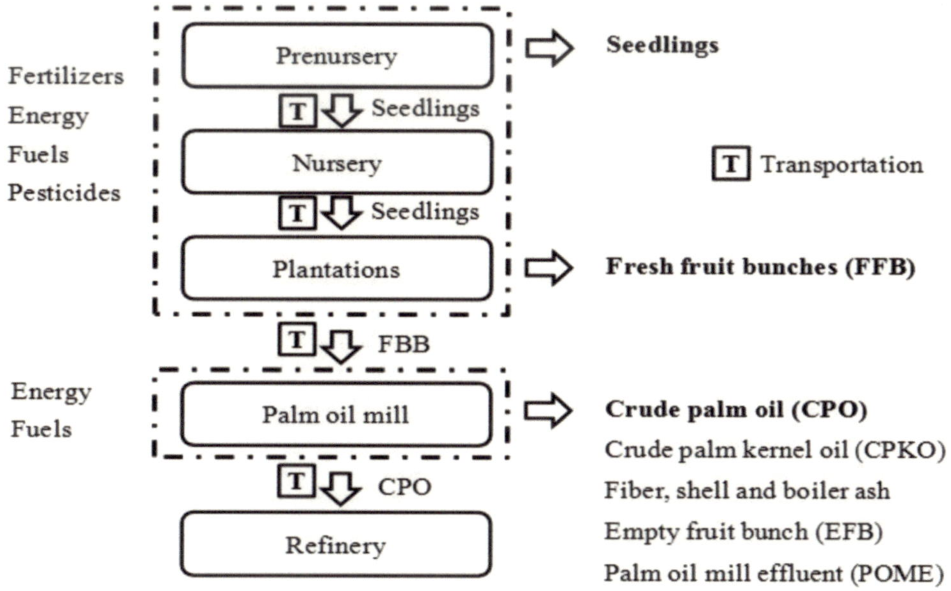

Figure 2.
Description of the CPO production system in the Brazilian Amazon region.

Subsequently, the CPO, the main product from the palm oil mill, is transported to a refinery aiming to produce refined palm oil (RPO). Although it is possible to use CPO as raw material, the biodiesel production is based on fatty acids extracted from palm oil fruit through the refining process [12].

2.2.1 *Agricultural phase*

In the oil palm cultivation, the first stage in the agricultural phase is the seedlings production. In prenursery, seeds are sown in small polyethylene bags (0.5 L) where the seedlings are kept under the shade to protect them from direct sunlight until they are 3 months old. In the subsequent main nursery stage, the seedlings are transferred to large polyethylene bags (18 L) and grown without a protective shade until they are 12 months old, when they are considered ready to be shifted to the plantations. Sprinkler systems are used to provide sufficient water in prenursery and nursery (8 mm day^{-1}). The seedlings are supplied with nutrients by fertilizer applications.

The palms from nursery are transferred to oil palm plantations and planted at a density of 143 plants/ha on a mineral soil with low and medium clay content. About some time after palm plantation, a legume used as cover crop (*Pueraria phaseoloides*) is sown. The cover crop prevents erosion and fixes nitrogen from the atmosphere in their root nodules, especially during the stage when the palms are young. A circle with no vegetation is established around each plant, preventing competition of weeds. The circle allows the herbicide application and the easy access for harvesting and picking of loose fruits.

The fertilizers applied in the oil palm plantations are potassium chloride, ammonium sulfate, kieserite, and rock phosphate. The herbicide glyphosate is used mainly to make the circles around the plant. The insecticide acephate is used in a small quantity due to use of integrated pest management, where natural enemies are used instead of pesticides.

The harvest operations start at 3 years of age and continue until the oil palm plantations are 25 years old. Harvesting of ripe FFB is manually carried out every 12 days using a sickle mounted on an aluminum pole. Normally, two fronds beneath the fruit bunch are pruned before harvesting. The pruned fronds are placed in the field between the palm rows for mulching. Detached FFBs are placed by the roadside, collected and disposed in dumpsters, and later taken by truck to a palm oil mill.

Replanting of oil palm is carried out after 25 years due to difficulty in harvesting tall palms and to low FFB yield. So, the palms are felled, chipped, and left in the plantation as a nutrient source for the new palm plants.

2.2.2 *CPO production*

In the palm oil mill, the FFBs are transferred into the sterilizer. The fruits are sterilized by steam (135°C) under a pressure of 3 kg cm^{-2} for 80–90 min. The sterilization process avoids loosens of the individual fruits from the bunch and also deactivates the enzyme which the breakdown of the oil into FFA.

The sterilized FFBs are sent to a thresher where the fruits are separated from the bunch. The empty fruit bunches (EFBs), which are abundant, may be used to produce steam and power, and the ashes used as fertilizers [13]. However, in the Agropalma farm, the EFBs also are applied on the organic plantations as organic fertilizer.

The fruits from the thresher are then sent to a digester that converts the fruits into a homogeneous oily mash by means of a mechanical stirring process. The digested mash is then pressed using a screw press to remove the major portion of the CPO. At this point, the CPO comprises a mixture of oil, water, and fruit solids, which are screened through a vibrating screen to remove as much solids as possible. The oil is then clarified in a continuous settling tank whose decanted CPO is then passed through a centrifugal purifier to remove remaining solids and then sent to the vacuum dryer to remove moisture. The CPO is then pumped to storage tanks before it is sent off to the refining process.

The nuts with the pressed mesocarp fibers are separated at the fiber cyclone and then cracked to produce kernels and shells. The kernels are shipped to kernel crushing plants to be processed into crude palm kernel oil (CPKO), while the shell and pressed mesocarp fiber are used as boiler fuel.

The main solid residues from the milling process are EFBs, pressed mesocarp fiber, shell, and boiler ash, while the liquid waste is palm oil mill effluent (POME) (**Figure 2**). In the studied plantations, the POME is conveyed from the mill and disposed in anaerobic ponds and later is used for palm tree irrigation purposes. All the residues from the palm oil extraction process are reused in the oil palm cultivation (EFB, ashes, and POME) or palm oil mill (fiber and shell).

2.2.3 Transportation of CPO from palm oil mill

The CPO stored in tanks is transported by trucks to the barge docks 24 km away from the mills and then is taken by barge to the refinery located 200 km from the farm. Each barge carries 1100 t of CPO. Besides being cheaper due to geographical conditions of the palm oil mill, the CPO transportation by the waterway is easier than it is by road in the region of study.

2.3 Scope definition and data collection

The scope of this study comprised since the stage of oil palm seedling production until the transportation of CPO. Inventory data included the main steps of the palm oil supply chain: agricultural production of FFB, extraction of CPO, and transportation of CPO to a refinery. We considered the year of 2009 (January 1 to December 31) to evaluate the carbon footprint of CPO production and transportation at Agropalma farm. In this study, the functional unit (FU) is 1 t of CPO.

We evaluated the GHG emissions related to seedlings production, planting and cultivation of juvenile and mature oil palm plantations, considering a FFB yield of 21.2 t ha^{-1}. The study measured direct and indirect emissions from the use of fuels, fertilizers (i.e., nitrogen, phosphorous, and potassium) and defensives in the production system.

The GHG emissions from industrial processes (by the use of fossil fuels, biofuels, biomass, and electricity) and FFB transportation from the field to the palm oil mill (by the use of fossil fuels) were calculated. In addition, we evaluated the GHG emissions from CPO transportation by trucks to the barge docks and then to the refinery. The GHG sources associated to inputs and outputs from the palm oil production, extraction, and transportation are described in **Figure 2**. The input data used to calculate the GHG emissions are listed in **Table 1**. **Table 2** presents the emission factors (EFs) used in this study.

The international standards ISO 14040 and ISO 14044 on life cycle assessments were used in this study to determine the GHG emissions from CPO production based in the methodology proposed by IPCC [18]. When allocation could not be avoided in the treatment of coproducts and residues, the resulting emissions of a process were portioned between its different products in a way that reflected the underlying physical relations between then. The rationale in using mass allocation is that physical portioning is most consistent as it contains the least uncertainties. According to the production at Agropalma in 2009, the weight allocation was 92% of CPO and 8% of PKO. To measure GHG emissions, we have used the inputs presented in the inventory data (**Table 1**) and their respective emission factors (**Table 2**). The EFB and POME are used in the plantations as fertilizers, while the

GHG source	Production stage		
	Agricultural	Industrial	Transportation
Nitrogen (N)	362	—	—
Phosphate (P)	256	—	—
Potassium (K)	3634	—	—
Pesticides (L)			
Glyphosate	52,745	—	—
Organophosphate	6678	—	—
Fossil fuel (L)[1]			
Diesel	3,708,216	854,458	365,909
Gasoline	124,834	64,476	—
Energy			
Electricity energy (kW/h)	—	1,685,636	—
Biomass (fiber + shell) (t)	—	80,064	—
Residue			
POME at anaerobic ponds (m^3)	—	693,193	—

[1]The quantities of biodiesel and ethanol mixed in diesel and gasoline, respectively, were computed.

Table 1.
Fertilizers, pesticides, fossil fuel and energy inputs, and waste generated in the crude palm oil (CPO) production in Pará State, Brazil.

Source	Unit	Emission factor	Reference
Fuel			
Diesel	kg CO_2/L diesel	3.11	[14]
Gasoline	kg CO_2/L gasoline	2.85	[14]
Ethanol	kg CO_2/L ethanol	0.46	[15]
Biodiesel	kg CO_2/L biodiesel	0.39	[15]
Fertilizer			
Nitrogen	kg CO_2/kg N	3.14	[16]
Phosphate	kg CO_2/kg P	0.61	[16]
Potassium	kg CO_2/kg K	0.44	[16]
Electricity			
2009 (medium value)	t CO_2/MWh	0.0245	[17]
Herbicide			
Glyphosate	kg CO_2 eq/kg A.I.[1]	15.95	[14]
Insecticide			
Organophosphate	kg CO_2 e/kg A.I.	7.68	[14]
Residues			
POME at anaerobic ponds	kg CO_2 e/kg DQO	5.00	[14]
Biomass (fiber + shell)	kg CO_2 e/t	22.53	[18]

[1]A.I., active ingredient.

Table 2.
Emission factors used to calculate the GHG emissions from crude palm oil (CPO) production in Pará State, Brazil.

pressed fiber and shell are burnt as fuel in the palm oil mill boiler. Nevertheless, the savings for the use of solid residues and POME are not included in this study.

Based on the input data, it was possible to estimate GHG emissions in terms of equivalent CO_2 (CO_2-eq). Emissions of N_2O and CH_4 are compared based on the global warming potential (GWP), since CH_4 and N_2O have a GWP 25 and 298 times larger than CO_2, respectively [18].

3. Results and discussion

3.1 Total GHG emissions

The total GHG emissions resulting from the production of FFB, extraction and transportation of CPO were 112,825 t CO_2-eq (**Figure 3**). The main source of GHG was the management of POME, followed by fertilizer application, fuel combustion, pesticide application, and electricity use. The high CH_4 emissions in the anaerobic ponds, when converted into CO_2-eq represented 66.5% of the total emissions. The use of fertilizers and fuels contributed to 17.9 and 15.1% of the total emissions, respectively. The application and electricity use represented less than 1% of the total GHG emissions (**Figure 4**).

The highest GHG amount emitted from POME is related to CH_4 emissions in the anaerobic ponds. The anaerobic ponds located in Agropalma farm are over 2 m deep and the POME has a large amount of C available and, consequently, high COD (chemical oxygen demand). In this study, the COD had an average of 21.65 kg m^{-3}, which is considered a high value by the standards of the Intergovernmental Panel on Climate Change [18] for effluents generated in the vegetable oil industry.

Similar results were found by Choo et al. [19] determining the GHG contribution by subsystems in the oil palm supply chain at Malaysia. The highest emissions were associated to POME without biogas capture. After POME disposal into the ponds, the CH_4 content is higher in the outlet region [20]. This is explained by excess concentration of organic matter in the inlet region that would influence the methanogenic activities [21]. As a result, lower CH_4 and higher CO_2 are emitted from the inlet region.

The fertilizers applied in the seedlings production and oil palm plantations resulted in the second largest GHG emission source. It is important to notice that GHG emissions are released both during the industrial production of fertilizer and through the application of these fertilizers on the field. The cultivation of oil palm requires high inputs of nitrogen fertilizer, which can create high soil emissions due to a conversion into N_2O [22].

The oil palm cultivation in Agropalma farm uses small amount of pesticides, resulting in low GHG emissions due to the use of these products (**Figure 3**). In 2009, glyphosate (herbicide) and acephate (insecticide) were used, so that GHG emissions were lower than other results previously reported in Malaysia [13, 19] and in Brazil [12].

The measured fuel combustion was related to agricultural operations, FFB transportation, extraction and transportation of CPO to a refinery. According to previous studies [13, 23], GHG emissions associated to use of diesel in plantations, internal transport and machinery are in order of 180–404 kg CO_2-eq ha^{-1} $yr.^{-1}$. If the FFB yield is set at 21.2 t ha^{-1}, the total emissions would be in order of 8.5–19.1 kg CO_2-eq/t FFB. We found a total emission of 23.08 kg CO_2-eq/t FFB and this value can be attributed to the highest number of mechanized operations at Agropalma farm compared with the most part of palm oil companies in Southeast Asia. Moreover, we also include the emissions from the combustion of fibers and

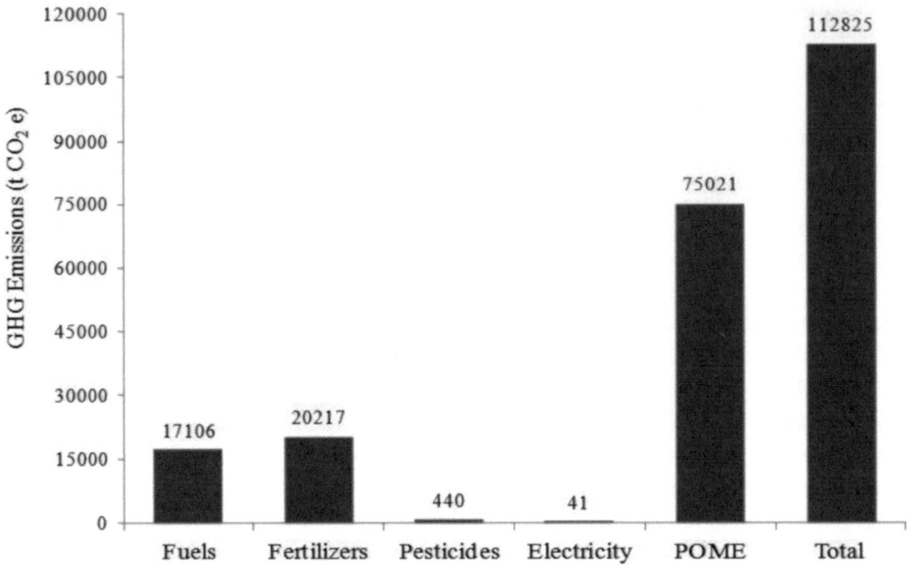

Figure 3.
Total GHG emissions (t CO_2-eq) in 2009 from crude palm oil production in the Brazilian Amazon region.

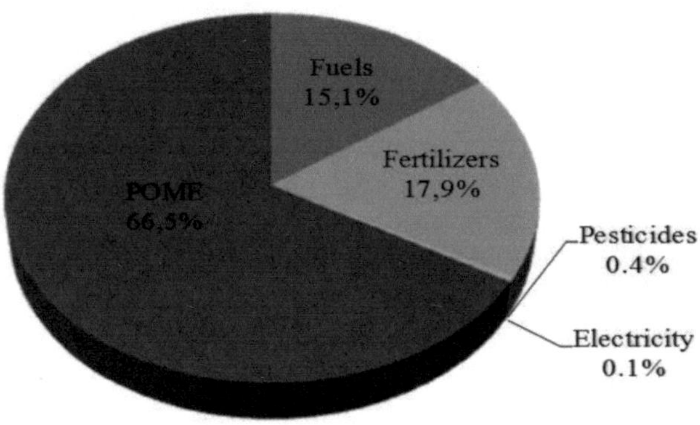

Figure 4.
Partition of total GHG emissions from crude palm oil production in the Brazilian Amazon region.

shells used in the boilers, which represent 10.5% of total emissions from the use of fuels.

The electricity is derived from hydroelectric power plants, which are considered a clean energy source. So, the GHG emissions were low, in agreement with previous results reported by Souza et al. [12]. The use of biomass (shell and fiber) in the boilers also contributed to the reduced use of electricity. In Malaysia, Yee et al. [24] have reported that from the amount of energy generated from the fibers and shell, about 55–77% is being utilized in the milling processes in the form of heat (steam) and power (electricity). The combustion of coproducts in high-efficiency boilers and turbines for power production reduces life cycle GHG emissions even when the most part of the electricity consumed comes from hydroelectric power plants [12].

3.2 GHG emissions in the agricultural phase

The agricultural phase (seedling production, juvenile and mature plantations) contributed to 28% of the total GHG emissions in the palm oil production, emitting

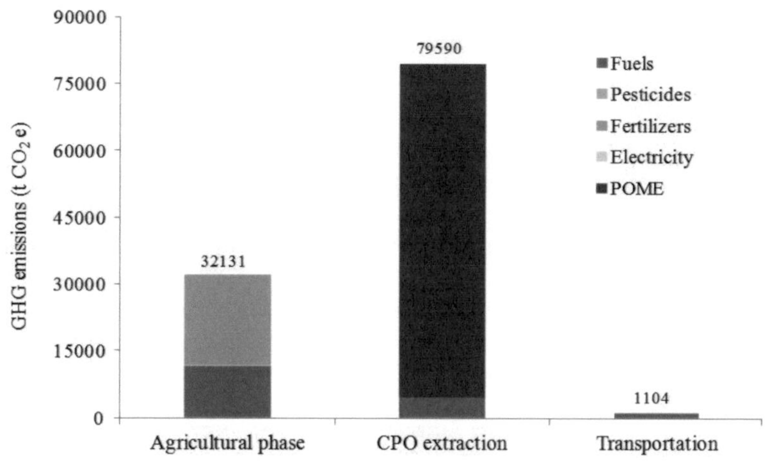

Figure 5.
GHG emissions (t CO_2-eq) from agricultural phase, CPO extraction and transportation in the Brazilian Amazon region.

32,131 t CO_2-eq in 2009 (**Figure 5**). The use of fertilizer was the main source of GHG emissions, followed by the use of fuels and pesticides. We found that in the production of FFB, 88% of GHG emissions are related to mature crop stage, while 1 and 9% are from the nursery and juvenile stands, respectively (**Figure 6**). These results are in agreement with other studies analyzing the GHG emission sources in the agricultural phase of oil palm production [23, 25].

In the oil palm nursery, the inputs of fertilizers and pesticides are relatively low compared to the other stages, and the activities for the seedlings cultivation are performed manually. The use of fuel is required only for the seedlings transportation and irrigation. So, this step contributed to insignificant amounts to the GHG emissions, corroborating with the results reported by Choo et al. [19].

In the oil palm plantations (juvenile and mature stands), the fertilizer application rate is dependent on a number of factors including yield potential, age of palm plant, nutrient balance, field conditions, and soil types. In 2009, the medium amount or

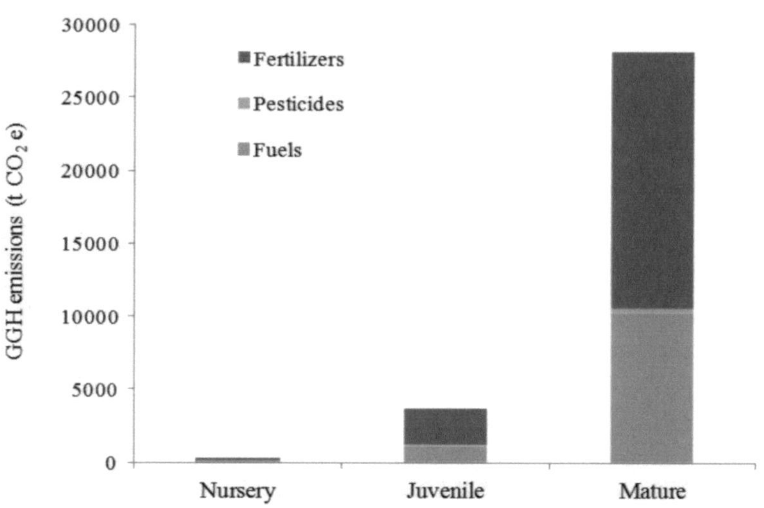

Figure 6.
GHG emissions (t CO_2-eq) in the agricultural phase of CPO production in the Brazilian Amazon region.

fertilizer applied for juvenile stands (1–3 years) was 36 kg N ha^{-1}, while in the mature stands (FFB production areas), it was 48 kg N ha^{-1}. We found that the use of fertilizer contributes to 63% of GHG emissions in the agricultural phase, agreeing with previous results reported by Yee et al. [24], Souza et al. [12], and Choo et al. [19].

The application of fertilizers and harvesting operations and the transportation of FFB from the field to the mill require the use of fossil fuels and biofuels. GHG emissions from the use of fossil fuels and biofuels accounted for 36% of total GHG emissions in the agricultural phase.

3.3 GHG emissions in the CPO extraction

In the palm oil milling stage, the GHG emissions were from the use of fuels, electricity, and disposal of POME. We observed that CPO extraction was the largest source of GHG emissions (71%) in 2009 (**Figure 5**).

As mentioned before, the use of fuels in the palm oil mill is represented by fossil fuels, biofuels, and residues of CPO extraction (fiber and shells). In the Agropalma farm, 80,000 t of fiber and shells were used as fuel in 2009. Normally, biomass is used for heat and/or power production through direct combustion. Yee et al. [24] have reported that palm oil mills in Malaysia are self-sufficient in terms of electricity consumption due the use of fiber and shells as source of power. The use of fossil fuels blended with biofuels, and residues of CPO extraction contributed to 3.4 and 2.3% of the total GHG emissions (**Figure 7**).

The CH_4 from POME in anaerobic ponds represented 94.3% of the total GHG emissions in the palm oil mills. In Malaysia, Shirai et al. [26] and Yacob et al. [27] have reported that the CH_4 composition was between 35 and 45% for the anaerobic treatment of POME, while Yacob et al. [20] recorded an average of 54.4%. After 1 year of observation at the anaerobic ponds, these authors observed that CH_4 emission pattern is governed by the oil palm seasonal cropping and mill activities. In this study, we used the default value proposed by IPCC, so the continuous monitoring is necessary to obtain the seasonal fluctuations in GHG emissions.

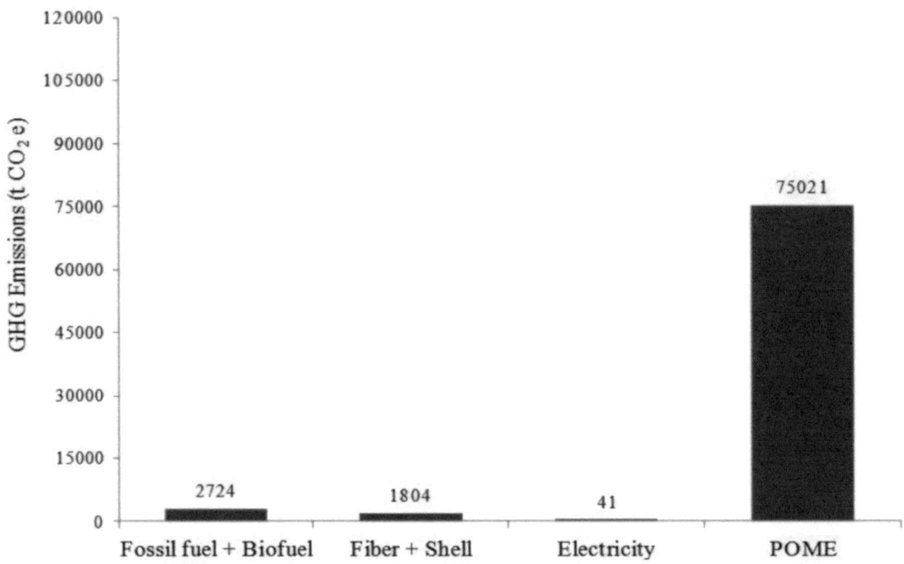

Figure 7.
GHG emissions (t CO_2-eq) in the palm oil milling stage in the Brazilian Amazon region.

Determining the GHG emissions by subsystems in the oil palm supply chain, Choo et al. [19] also reported the highest emissions associated to POME production. Their results showed reduction of GHG emissions from oil palm mill when the CH_4 was captured. Reijnders and Huijbregts [23] reported a reduction on GHG emissions of about 0.15 t CO_2-eq/t CPO produced when 95% of control efficiency for the treatment of POME is assumed.

3.4 GHG emissions from the CPO transportation

The fuel combustion during the CPO transportation from the palm oil mill to the refinery represented 1% of the total GHG emissions in 2009 (**Figure 5**). As mentioned before, the transportation of CPO from oil palm mill to refinery at Agropalma farm is performed by barge where one roundtrip carries 1100 t of CPO and consumes 3500 L of diesel. Since 105 roundtrips were performed in 2009, the GHG emissions from the use of diesel at Agropalma farm during the transportation of CPO were 1104 t CO_2-eq.

According to Majer et al. [22], the GHG emissions from transportation process typically represent only a small share of the overall balance results in the biodiesel production. In contrast to our results, Choo et al. [19] reported higher GHG emissions from the transportation process in Malaysia due the distance of oil palm mill until the refinery. The authors point out those GHG emissions could be reduced by improving transport logistic by routing delivery of CPO, for the shortest distance between the supplier and the refinery.

3.5 Carbon footprint of CPO production and transportation

In the palm oil mill, CPO and KPO are obtained as main products among the several by-products from the extraction process (**Figure 2**). Several studies have reported the GHG emissions considering the production area or the quantity produced [12, 19, 23, 28]. We calculated the GHG emissions related to CPO production using mass allocation based on a specific agricultural year. The carbon footprint was calculated considering the GHG emissions per t of CPO produced at Agropalma farm. In 2009, the CPO and KPO production were 130,210 and 11,205 t, respectively.

Considering the agricultural production of FFB, CPO extraction, and CPO transportation from the mill to the refinery, we found an emission of 0.79 t CO_2-eq/t CPO produced at Agropalma farm in 2009. As we mentioned before, 66.5% of the total emissions are related to management of POME in the anaerobic ponds. The C footprint of CPO can be reduced significantly since oil palm and palm oil processing wastes are used to replace the input of fossil fuel in palm oil processing stage [29]. This can be combined with a reduction in the amount of CH_4 emitted from oil palm processing waste [23]. The company could adopt the system of CH_4 capture from POME and use it as electricity or power source [19, 20].

3.6 Comparative GHG emissions in CPO production

Our study is the first to approach carbon footprint considering the different stages (agricultural, industry, and transportation) of CPO production in Brazil. We found that industry is the main source of GHG emissions (71%) due the management of POME in anaerobic ponds. Another study in Thailand [6] has reported similar results, and the GHG emissions in CPO extraction (industry) allocated by energy value were 0.55 t CO_2-eq/t CPO. But contrary to what was observed in this study, the authors have considered the carbon stocks.

Determining the GHG contributions by subsystems in the oil palm supply chain using the LCA approach in Malaysia, Choo et al. [19] have reported that the production of 1 t of CPO in a mill without and with biogas capture emitted 0.97 and 0.51 t CO_2-eq, respectively. As we found is this study, the contribution of nursery subsystem was found to be minimal, and in the plantation subsystem the major sources of GHG were from nitrogen fertilizers.

Regarding the soil GHG emissions in Indonesian oil palm plantations, Rahman et al. [5] have reported that the use of inorganic fertilizers led to significantly higher N_2O emissions. Therefore, as we found is this study that the use of fertilizers accounted for 63% of GHG emissions in the agricultural phase, the use of organic amendments (empty fruit bunches, enriched mulch, and pruned oil palm fronds) can be an option for reducing GHG emissions.

3.7 Mitigation of GHG emissions and opportunities do achieve CO_2 sequestration

The GHG emissions have been reported along the palm oil production chain from the roundtable on sustainable palm oil (RSPO) [30]. Methane (CH_4) emissions from wastewater in open ponds at the milling phase and nitrous oxide (N_2O) emissions from nitrogen fertilizer application in the cultivation phase are the most related sources of GHG emissions [31, 32].

In this study, the C footprint associated to CPO production was about 0.79 kg CO_2/kg CPO and the main source of GHG emissions is associated to management of POME in the anaerobic ponds. Previous study in Thailand also has reported that wastewater treatment and empty-fruit-bunch disposal in mills are a main source of CH_4 emissions and cause global warming, with up to 47 and 45% of total global warming impact [33]. So, the effluent treatment in the anaerobic ponds and the combustion of CH_4 during anaerobic decomposition [34] are cited as viable strategies to reduce the GHG emissions at the milling phase.

According to Chai et al. [35], energy content in wastewater, in the form of chemical oxygen demand (COD), is usually converted into CO_2 or CH_4 and biosolids through either aerobic treatment or anaerobic treatment. Therefore, decreasing the degree of aerobic treatment and maximizing energy recovery from CH_4 and biosolids are crucial to lower carbon footprint. An efficient anaerobic digestion could contribute to the decrease of the degree of subsequent aerobic treatment, by removing certain amount of COD and reducing CO_2 emissions, and recover energy from CH_4 by anaerobic digestion.

When the POME is converted into biogas (CH_4) through a gasification process and then used to fuel gas engine and generate electricity, it is possible to reduce the environmental impact of CPO production. In addition, other recent technological advances have turned POME to useful sustainable feedstock that can be used to produce valuable by-products like biohydrogen [36] and biomethane [37].

Regarding the oil palm plantations, there are four main steps that contribute to GHG emissions: soil preparation, fertilizer management, weed control, and FFB transportation. In this study, we reported that the use of fertilizer was the main source of GHG emissions, followed by the use of fuels and pesticides. The use of EFB for infield application in young and mature palm areas has been used in the management of soil nutrients and organic matter and promoting the increase of organic carbon in the soil over time [38, 39]. So, the continuous use of EFB as mulching could play a significant role in reducing CO_2 emissions into the atmosphere through soil C sequestration [39].

As also mentioned before, the use of *Pueraria phaseoloides* as cover crop can also reduce the use of nitrogen fertilizers in the young and mature palm plantations.

Previous study has reported that the use of legumes contributes about 150 kg nitrogen ha^{-1} $year^{-1}$ to the system through biological nitrogen fixation [40]. So, the use of EFB and legumes in the agricultural phase can reduce the mineral fertilizer demand and consequently minimize GHG emissions.

In the Brazilian Amazon region, the maintenance of native vegetation and the use of degraded areas to introduce new oil palm plantations can promote environmental benefits to commercial farms, reducing the GHG emissions. Brazilian government approved a bill to expand 4.3 million ha of previously deforested lands to oil palm plantations [41]. Pará State is intensively studied because the majority of the land deemed suitable for oil palm expansion by the government is located in this state. According to Yui and Yeh [42], the encouragement of oil palm plantations on deforested lands could drastically reduce the conversion of forest land, thus reducing GHG emissions from deforestation.

4. Conclusions

Considering the production of seedlings and FFB, the extraction and transportation of CPO in a Brazilian commercial farm at Amazon region, it emitted 0.79 t CO_2-eq/t CPO produced in 2009. Main contributing factor to GHG emissions during the cultivation step is the use of industrial fertilizers, which accounted for 17.9% of the total GHG emissions mainly due the high input of nitrogen. The management of POME from palm oil mill is the main source of GHG emissions to the atmosphere, representing 66.5% of the carbon footprint during the evaluated period. Regarding the use of fuels in all evaluated stages, they accounted for 15.1% of the total GHG emissions.

The POME treatment in the anaerobic ponds and the use of CH_4 for steam or electricity production and the use of EFB and legumes in the agricultural phase are cited as the main strategies to reduce the GHG emissions in the palm oil production system.

Our results may be used to encourage new researches and improve the life cycle assessment and the measurements of GHG emissions associated to palm oil production chain in Brazil and other regions of South America.

Author details

Leidivan Almeida Frazão[1]*, Guilherme Silva Raucci[2], João Luis Nunes Carvalho[3], Marcelo Valadares Galdos[4], Cindy Silva Moreira[5], Carlos Eduardo Pellegrino Cerri[6] and Carlos Clemente Cerri[6]

1 Federal University of Minas Gerais, Montes Claros, MG, Brazil

2 Agrosmart, Campinas, SP, Brazil

3 Brazilian Biorenewables National Laboratory, Campinas, SP, Brazil

4 University of Leeds, Leeds, UK

5 Abiove, São Paulo, SP, Brazil

6 University of São Paulo, Piracicaba, SP, Brazil

*Address all correspondence to: lafrazao@ufmg.br

References

[1] Barcelos E, Almeida Rios S, Cunha RNV, Lopes R, Motoike SY, Babiychuk E, et al. Oil palm natural diversity and the potential for yield improvement. Frontiers in Plant Science. 2015;**6**:190

[2] Ministério da Agricultura, Pecuária e Abastecimento. Diagnóstico da produção sustentável do óleo de palma. Brasília: MAPA/ACE; 2018

[3] Empresa Brasileira de Pesquisa Agropecuária. Dendê. Brasil: Ministério da Agricultura, Pecuária e Abastecimiento; 2010

[4] Glass V. Expansão do dendê na Amazônia brasileira: Elementos para uma análise dos impactos sobre a agricultura familiar no nordeste do Pará. Brasil: Centro de Monitoramento de Agrocombustíveis. ONG Repórter; 2013

[5] Rahman N, Bruun TB, Giller KE, Magid J, Ven GWJ, de Neergaard A. Soil greenhouse gas emissions from inorganic fertilizers and recycled oil palm waste products from Indonesian oil palm plantations. GCB Bioenergy. 2019;**11**:1056-1074

[6] Bunchai A, Suttinun O, H-Kittikun A, Musikavong C. Life cycle greenhouse gas emissions of palm oil production by wet and dry extraction processes in Thailand. International Journal of Life Cycle Assessment. 2017;**22**:1802-1814

[7] Menichetti E, Otto M. Energy balance & greenhouse gas emissions of biofuels from a life cycle perspective. In: Howarth RH, Bringezu S, editors. Proceedings of the Scientific Committee on Problems of the Environment (SCOPE) International Biofuels Project Rapid Assessment. Gummersbach; 2009

[8] Empresa Brasileira de Pesquisa Agropecuária. Sistema Brasileiro de Classificação de Solos. 2nd ed. Rio de Janeiro: Centro Nacional de Pesquisa de Solos; 2013

[9] United States Department of Agriculture. Soil Taxonomy: A Basic System of Soil Classification for Making and Interpreting Soil Surveys. 3rd ed. Washington: Soil Survey Staff; 1999

[10] Teoh CH. The Palm Oil Industry in Malaysia: From Seed to Frying Pan. Malaysia: WWF; 2002

[11] Arnold MG, Teoh KT, Carlin G. Steam (physical) refining deodorizer for Malaysian palm oil. Journal of the American Oil Chemists Society. 1997;**54**:312-316

[12] Souza SP, Pacca S, Ávilla MT, Borges JL. Greenhouse gas emissions and energy balance of palm oil biofuel. Renewable Energy. 2010;**35**:2552-2561

[13] Wicke B, Dornburg V, Junginge RM, Faaij M. Different palm oil production systems for energy purposes and their greenhouse gas implications. Biomass and Bioenergy. 2008;**32**:1322-1337

[14] Centre E. Ecoinvent Data. v2.2. Dübendorf: Swiss Centre for Life Cycle Inventories; 2010

[15] Macedo IC, Seabra JEA, Silva JEAR. Greenhouse gases emissions in the production and use of ethanol from sugarcane in Brazil: The 2005/2006 averages and a prediction for 2020. Biomass and Bioenergy. 2008;**32**:582-595

[16] West T, Marland GA. Synthesis of carbon sequestration, carbon emissions, and net carbon flux in agriculture: Comparing tillage practices in the United States. Agriculture, Ecosystems & Environment. 2002;**91**:217-232

[17] Ministério da Ciência e Tecnologia—MCT. Fatores de Emissão de CO_2 para

utilizações que necessitam do fator médio de emissão do Sistema Interligado Nacional do Brasil, como, por exemplo, inventários corporativos. Brasília: MCT; 2010

[18] Intergovernmental Panel on Climate Change. Guidelines for National Greenhouse Gas Inventories: Agriculture, Forestry and Other Land Use. Vol. 4. Hayama: National Greenhouse Gas Inventories Programme; 2006

[19] Choo YM, Muhamad H, Hashim Z, Subramaniam V, Puah CW, Tan YA. Determination of GHG contributions by subsystems in the oil palm supply chain using the LCA approach. International Journal of Life Cycle Assessment. 2011;**16**:669-681

[20] Yacob S, Hassan MA, Shirai Y, Wakisava M, Subash S. Baseline study of methane emission from anaerobic ponds of palm oil mill effluent treatment. Science of the Total Environment. 2006;**366**:187-196

[21] Masse L, Massé DI. Effect of soluble organic, particulate organic, and hydraulic shock loads on anaerobic sequencing batch reactors treating slaughterhouse wastewater at 20 C. Process Biochemistry. 2005;**40**:1225-1232

[22] Majer S, Mueller-Langer F, Zeller V, Kaltschmitt M. Implications of biodiesel production and utilization on global climate – A literature review. European Journal of Lipid Science and Technology. 2009;**11**:747-762

[23] Reijnders L, Huijbregts MAJ. Palm oil and the emission of carbon-based greenhouse gases. Journal of Cleaner Production. 2008;**16**:477-482

[24] Yee KF, Tan KT, Abdullah AZ, Lee KT. Life cycle assessment of palm biodiesel: Revealing facts and benefits for sustainability. Applied Energy. 2009;**86**:189-196

[25] Hardter R, Chow WY, Hock OS. Intensive plantation cropping: A source of sustainable food and energy production in the tropical rainforest areas in Southeast Asia. Forest Ecology and Management. 1997;**93**:95-102

[26] Shirai Y, Wakisaka M, Yacob S, Hassan MA, Suzuki S. Reduction of methane released from palm oil mill lagoon in Malaysia and its countermeasures. Mitigation and Adaptation Strategies for Global Change. 2003;**8**:237-252

[27] Yacob S, Hassan MA, Shirai Y, Wakisava M, Subash S. Baseline study of methane emission from open digesting tanks of palm oil mill effluent treatment. Chemosphere. 2005;**59**:1575-1581

[28] Stichnothe H, Schuchardt F. Life cycle assessment of two palm oil production systems. Biomass and Bioenergy. 2011;**35**:3976-3984

[29] Yusoff S. Renewable energy from palm oil e an innovation on effective utilization of waste. Journal of Cleaner Production. 2006;**14**:87-93

[30] Bessou C, Chase LC, Henson IE, Abdul-Manan AFN, Milà I, Canals L, et al. Pilot application of palm GHG, the roundtable on sustainable palm oil greenhouse gas calculator for oil palm products. Journal of Cleaner Production. 2014;**73**:136-145

[31] Silalertruksa T, Gheewala SH. Sustainability assessment of palm biodiesel production in Thailand. Journal of Energy. 2012;**43**:306-314

[32] Kaewmai R, H-Kittikun A, Musikavong C. Greenhouse gas emissions of palm oil mills in Thailand. International Journal of Greenhouse Gas Control. 2012;**11**:141-151

[33] Saswattecha K, Kroeze C, Jawjit W, Hein L. Assessing the environmental

impact of palm oil produced in Thailand. Journal of Cleaner Production. 2015;**100**:150-169

[34] Rana S, Lakhveer Singh L, Wahid Z, Liu H. A recent overview of palm oil mill effluent management via bioreactor configurations. Current Pollution Reports. 2017;**3**:254-267

[35] Chai C, Zhang D, Yu Y, Feng Y, Wong MS. Carbon footprint analyses of mainstream wastewater treatment technologies under different sludge treatment scenarios in China. Water. 2015;**7**:918-938

[36] Sivasangar S, Zainal Z, Salmiaton A, Taufiq-Yap YH. Supercritical water gasification of empty fruit bunches from oil palm for hydrogen production. Fuel. 2015;**143**:563-569

[37] Montafia P, Gnansounou E. Life cycle assessment of thermochemical conversion of empty fruit bunch of oil palm to biomethane. In: Edgard Gnansounou E, Pandey A, editors. Life-Cycle Assessment of Biorefineries. Amsterdam: Elsevier; 2017

[38] Zaharah AR, Lim KC. Oil palm empty fruit bunch as a source of nutrients and soil ameliorant in oil palm plantation. Malaysian Journal of Soil Science. 2000;**4**:51-66

[39] Moradi A, Sung CTB, Joo GK, Hanif AHN, Ishak CF. Soil organic C sequestration due to different oil palm residue mulches. Advances in Tropical Soil Science. 2013:169-186

[40] Agamuthu P, Broughton WJ. Nutrient cycling within the developing oil palm-legume ecosystem. Agriculture, Ecosystems and Environment. 1985;**13**:111-123

[41] Osava M. Brazil: Oil Palm Plantations Expand on Degraded Lands in Amazon. Rome: Inter Press Service News Agency; 2010. Available from: http://www.ipsnews.net/2010/12/brazil-oil-palm-plantations-expand-on-degraded-land-in-amazon/

[42] Yui S, Yeh S. Land use change emissions from oil palm expansion in Pará, Brazil depend on proper policy enforcement on deforested lands. Environmental Research Letters. 2013;**8**:1-9

Chapter 7

Experimental Study of Adsorption on Activated Carbon for CO_2 Capture

Hesham G. Ibrahim and Mohamed A. Al-Meshragi

Abstract

The adsorption of carbon dioxide (CO_2) on activated carbon (AC) prepared from olive trees has been investigated by using a fixed bed adsorption apparatus. The adsorption equilibrium and breakthrough curves were determined at different temperatures 30, 50, 70, and 90°C in order to investigate both kinetic and thermodynamic parameters. Maximum CO_2 sorption capacity on AC ranged from 109.5 to 35.46 and from 129.65 to 35.55 mg CO_2/g of AC for initial concentrations 10 and 13.725% vol., respectively. Different isotherm models are applied to mathematically model the CO_2 adsorption, and on the basis of the estimated adsorption capacity by model and determination coefficient (r^2), the Langmuir model provides a perfect fit to the experimental data owing to closeness of the r^2 to unity. From the correlation coefficient, it is found that the pseudo-second-order model is well-fitted with the experimental data. In addition, it indicates that CO_2 adsorption is a physical adsorption process and demonstrates a behavior of an exothermic reaction, which is consistent with the thermodynamic analysis. The results obtained in this study conclude that AC prepared from olive trees can be considered as adequate for designing a fixed bed cycle to separate carbon dioxide from flue gases and serve as a benchmark while searching for inexpensive and superior activated carbon production in future studies.

Keywords: adsorption, breakthrough, equilibrium, kinetic, thermodynamic

1. Introduction

The emissions of CO_2 from burn fossil fuels are the major reason for the increase in the concentration of this gas in the atmosphere [1]. The amount of carbon dioxide in the atmosphere is currently increasing globally by around 6 billion tons per year [2].

A feasible technique method used industrially in reduction of CO_2 emissions is capture and storage. CO_2 capture means separating the CO_2 from other gases in flue. The advanced technologies being used worldwide for CO_2 capture in different arrangements are post-combustion, pre-combustion, and oxy-fuel processes [1].

Numerous investigations have been done for CO_2 capture field by using adsorption, which are indicating to the effective usage of a post-combustion treatment of gas emissions of flue. The proposed schemes in a cycle process of capture by adsorption include pressure swing adsorption (PSA) and temperature swing adsorption (TSA) [1, 3–5]. The capture of carbon dioxide by adsorptive process is mainly based on preferential adsorption of this gas on a porous adsorbent. Thus, the first and most important step is to find a suitable adsorbent [1]. Carbon materials are relatively insensitive to moisture and are suitable candidates for CO_2 capture due to their pore structure and surface chemistry properties [6].

In recent years, considerable attention has been focused on removal of pollutants by using adsorbents derived from low-cost agro-wastes. Olive trees (*Olea europaea*) are abundantly found and easily available in the Mediterranean countries generally and especially Libya. Thus, the aim of the present study is to describe the dynamics and equilibrium of CO_2-N_2 mixture adsorption on local activated carbon (AC) prepared from olive trees using the breakthrough curve method. Experimental breakthrough curves are used to obtain equilibrium data, and then Langmuir, Freundlich, Temkin, and Dubinin-Radushkevich equilibrium adsorption models were applied. Kinetic models examined herein are simple first-order, pseudo-first-order, pseudo-second-order, and intra-particle diffusion. Model validity with experimental data is assessed by using the coefficient of determination (r^2); the closer the value to unity means that the model will be better. Thermodynamic analysis of adsorption of CO_2 on AC estimates the values of enthalpy, free energy, and entropy. Also, effects of the interaction between CO_2 and NO are studied.

2. Materials and methods

2.1 Preparation of activated carbon

The prepared activated carbon based on charcoal was prepared from olive trees for low cost and was abundantly available. The used activated carbon was obtained from the local area. The raw material of charcoal as received was crushed, ground, and sieved, and only the fraction of particle size 5 mm was chosen as the mean particle diameter. Then it is heated in an oven for 48 h up to 115°C to dry and activate (to remove the absorbed gases and moisture it contains) [7, 8]. The produced activated carbon is then stored in a tightly closed container to be used as required. The total pore volume and surface area of AC were determined using Gemini VII 2390a analyzer. The particle size is obtained by using standard mesh sieves (standard sieve AS 200), and average value of bed porosity is calculated in terms of the average diameter of particles [9].

2.2 Dynamic adsorption capacity of carbon dioxide

A laboratory system used for measuring breakthrough curve was set up and shown in **Figure 1**. The adsorber which is made of carbon steel tube, consists of three zones:

- Calming zone with 6.5 cm diameter and 8 cm length containing spherical particles of carbon steel.

- Active zone with 8.44 cm diameter and 39 cm length containing the activated carbon particles, and it was surrounded by a shell containing a heating medium.

- Ending zone with 6.5 cm diameter and 8 cm length containing spherical particles of carbon steel.

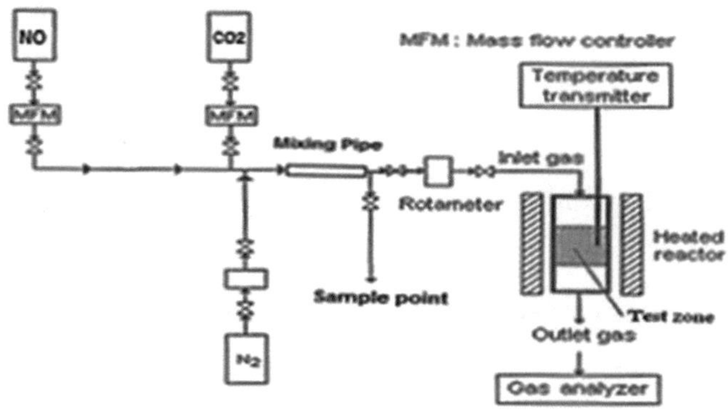

Figure 1.
Schematic of the experimental set-up.

N_2 and CO_2 were supplied by pressurized cylinders. The purity for CO_2, NO, and N_2 cylinders was 99.9, 99, and 99.99% (vol.%), respectively. The used concentrations of CO_2 were 10 and 13.725% (vol.%). Delivery of the feed gas was controlled by mass flow meter. After mixing in a mixing chamber (2.45 cm diameter and 15 cm length), simulated gas was fed into the inlet of the adsorber. Prior to all measurements, an initial degassing of the sample was performed at a given temperature (30, 50, 70, and 90°C) by the flow of nitrogen until reaching steady state. Then mixed gas was passed through the fixed bed column at constant temperature. The inlet and outlet concentrations were analyzed by a Testo 350XL flue gas, which has a resolution for N_2, NO, and CO_2 of 0.1 ppm, 0.1 ppm, and 0.1% vol., respectively. The total flow was kept constant for 12 l/min; whereas the N_2 and CO_2 were controlled precisely according to the required balance gas N2 during binary experiments. The dynamic adsorption capacity of CO_2 onto AC column was calculated using Eqs. (1) and (2) [10]:

$$t = \int_0^t \left(1 - \frac{C}{C_o}\right) dt \tag{1}$$

$$q_i = \frac{Q_v . y_i . t . \rho_i}{m} * 1000 \tag{2}$$

where t is the time of adsorption (min.), C is the outlet concentration of CO_2 gas (mg/l), C_o is the inlet concentration of CO_2 gas (mg/l), q_i is the amount of adsorbed gases (mg gas/g adsorbent), Q_v is the volumetric flow rate of CO_2 gas (l/min.), y_i is the mole fraction of inlet CO_2, ρ_i is the density of inlet gas (mg/l), and m is the weight of the adsorbent (g).

The interval times for measurements were 5, 10, 15, 20, 25, 30, 35, 40, 50, 60, 70, and 80 min. The experimental procedures and measurements are replicated three times for accuracy.

2.3 Adsorption isotherm studies

In order to optimize the design of a sorption system to capture CO_2 on AC, the suitable isotherm model for equilibrium curves must be established. Equilibrium models that have been examined herein are Langmuir, Freundlich, Temkin, and Dubinin-Radushkevich. The conformity between the predicted values of models and experimental data is expressed by comparing the experimental adsorption capacity with the adsorption capacity estimated by these models, by means of the determination coefficient (r^2, values close or equal to 1) [11, 12].

2.3.1 The Langmuir isotherm

The widely used Langmuir isotherm found as a successful application in many real sorption processes [12] is expressed as

$$q_e = \frac{K_L C_e}{1 + a_L C_e} \tag{3}$$

A linear form of this expression is

$$\frac{C_e}{q_e} = \frac{1}{K_L} + \frac{a_L}{K_L} C_e \tag{4}$$

where q_e is the amount of adsorbed CO_2 per unit weight of AC at equilibrium (mg/g) and C_e is the unadsorbed CO_2 concentration in effluent at equilibrium (mg/l). K_L is the Langmuir equilibrium constant, and K_L/a_L value is used to estimate the theoretical monolayer capacity of AC, Q_o (mg/g). Therefore, the plot of C_e/q_e versus C_e enables one to determine the constants a_L and K_L.

2.3.2 The Freundlich isotherm

The well-known Freundlich isotherm is often used for heterogeneous surface energy systems [12]. The Freundlich equation is given as

$$q_e = K_F C_e^{1/n} \tag{5}$$

A linear form of this expression is

$$\log q_e = \log K_F + \frac{1}{n} \log C_e \tag{6}$$

where K_F is the Freundlich constant (mg/g) and n is the Freundlich exponent. K_F and n can be determined from the linear plot of log q_e versus log C_e.

2.3.3 Temkin isotherm

The Temkin isotherm [13, 14] has been used in the following form:

$$q_e = \frac{RT}{b} \ln (A C_e) \tag{7}$$

A linear form of the Temkin isotherm can be expressed as

$$q_e = \frac{RT}{b} \ln A + \frac{RT}{b} \ln C_e \tag{8}$$

where A is the Temkin isotherm equilibrium binding constant (l/g), b is the Temkin isotherm constant, R is the universal gas constant (8.314 J/mol.K), T is the temperature, and B (=RT/b) is the constant related to heat of adsorption (J/mol).

The sorption data can be analyzed according to Eq. (8). Therefore, the plot of q_e versus ln(C_e) enables one to determine the constants A and B.

2.3.4 Dubinin-Radushkevich isotherm

The Dubinin-Radushkevich equation in Eq. (9) is as follows [15]:

$$q_e = q_m\, e^{-\beta\varepsilon^2} \tag{9}$$

A linear form of Dubinin-Radushkevich isotherm is

$$\ln q_e = \ln q_m - \beta\varepsilon^2 \tag{10}$$

where q_m is the Dubinin-Radushkevich monolayer capacity (mg/g), β is the Dubinin-Radushkevich isotherm constant (mol^2/kJ^2), and ε is the Polanyi potential, and it's related with equilibrium concentration as follows:

$$\varepsilon = \frac{RT}{M}\ln\left(1 + \frac{1}{C_e}\right) \tag{11}$$

where R is the universal constant of gases (8.314 J/mol.K), T is the experiment temperature (K), and M is the molecular weight of CO_2. The constant β gives the mean free energy of adsorption (E) for CO_2 molecules transported from the gas bulk to the surface of AC which is calculated by using Eq. (12) [14, 16]:

$$E = \frac{1}{\sqrt{2\beta}} \tag{12}$$

2.4 Kinetic models of adsorption

To determine an appropriate kinetic model is necessary to analyze the experimental data to investigate the mechanism of adsorption process that may include mass transfer or chemical reaction. Also, other extensive models applied to many models such as homogenous surface diffusion model and heterogeneous diffusion model (also known as pore and diffusion models, respectively) have been extensively applied to expound the adsorbate transfer onto the particles of adsorbent [17–19]. The determination coefficient (r^2) is used to examine the confirmation of the predicted values of models with experimental data (determination coefficient value close or equal to 1). The validity of these models is evaluated by the determination coefficient (r^2), which is within the range of 0–1, in which r^2 closer to unity implies the best fitting toward the particular kinetic model [20].

2.4.1 Simple first-order model

The sorption kinetic may be described by a simple order equation [21, 22]. The following simple first-order equation describes the change in bulk concentration:

$$C_t = C_o e^{k_1 t} \tag{13}$$

that can be rearranged to obtain a linear form

$$\log C_t = \frac{k_1}{2.303} t + \log C_o \tag{14}$$

where C_t and C_o are the concentration of adsorbate at time t and initially (mg/l), respectively, and k_1 is the first-order rate constant, (1/min).

Furthermore, Sparks [23] and Hossain et al. [21] proposed that the simple kinetic models such as first- or second-order rate equations are not applicable to the adsorption system with solid surfaces.

2.4.2 Pseudo-first-order model

The sorption kinetics may be described by pseudo-first Eq. (15) [13, 21, 24–26]:

$$\frac{dq_t}{dt} = k_1(q_e - q_t) \tag{15}$$

Integration of Eq. (15) and using the initial conditions $q_t = 0$ at $t = 0$ and $q_t = q_t$ at $t = t$ yield

$$\log\left(\frac{q_e}{q_e - q_t}\right) = \frac{k_1}{2.303}t \tag{16}$$

By rearrangement of Eq. (16), a linear form is obtained:

$$\log(q_e - q) = \log q_e - \frac{k_1}{2.303}t \tag{17}$$

where q_e is the amount of CO_2 adsorbed at equilibrium (mg/g), q is the amount of CO_2 adsorbed at time t (mg/g), and k_1 is the pseudo-first-order constant (1/min).

The pseudo-first-order constant k_1 and equilibrium adsorption q_e are determined by plot of $\log(q_e\text{-}q)$ versus t.

2.4.3 Pseudo-second-order model

The adsorption kinetics may also be described by pseudo-second-order Eq. (17) [13, 26–30]:

$$\frac{dq_t}{dt} = k_2(q_e - q_t)^2 \tag{18}$$

Integrating Eq. (18) and applying the initial boundaries yield

$$\frac{1}{(q_e - q_t)} = \frac{1}{q_e} + k_2 t \tag{19}$$

By rearrangement Eq. (19), a linear form is obtained:

$$\frac{t}{q_t} = \frac{1}{k_2 {q_e}^2} + \frac{1}{q_e}t \tag{20}$$

where k_2 is the equilibrium rate constant of pseudo-second-order adsorption (g/mg.min).

The slopes and intercepts of plots t/q_e versus t are used to calculate the pseudo-second-order rate constants k_2 and q_e.

2.4.4 Intra-particle diffusion model

The intra-particle diffusion model is expressed as [31–33]

$$q_t = k_p t^{0.5} + c \quad (21)$$

where k_p is a rate factor (present CO_2 adsorbed per minute). The plot of this model is multi-linear that indicates there are two or more steps occurring consecutively. The external surface/instantaneous adsorption stage occurred first in sharp portion. Then a gradual adsorption stage is in the second portion, where the controlled rate is the intra-particle diffusion. Final equilibrium stage occurs where intra-particle diffusion begins to slow down because of extremely low adsorbate concentrations in the bulk [24, 34].

2.5 Thermodynamic studies

Thermodynamic parameters were estimated from Langmuir isotherms by using the Van't Hoff's equation as in Eqs. (22) and (23). The thermodynamic parameters can be estimated from Langmuir isotherms by using the Van't Hoff's equation as follows [12, 35]:

$$\Delta G^o = -RT\ln a_L \quad (22)$$

$$\ln a_L = \frac{\Delta S^o}{R} - \frac{\Delta H^o}{RT} \quad (23)$$

where a_L is a Langmuir constant (l/mol), R is the universal constant of gases (8.314 J/mol.K), and T is an absolute temperature of gas.

3. Results and discussion

3.1 Adsorbent characterization

The main characteristics of AC (particle diameter, bed porosity, weight of bed, BET surface area, and pore volume) are shown in **Table 1**. Due to a high value BET surface area for used AC, its good pore structure makes it a suitable candidate for CO_2 capture.

3.2 Dynamic studies

Two mixtures of CO_2 and N_2 gases have been used in experiments (initial concentrations of CO_2 are 10 and 13.725% vol., respectively). **Figure 2** shows that

Characteristic	Value	Unit
Particle diameter	5	mm
Bed porosity	0.304	—
Weight of bed	500	g
BET surface area	602	m^2/g
Pore volume	0.61	cm^3/g

Table 1.
Characteristics of used AC depending on particle diameter.

the rate of CO_2 adsorption gradually decreased with time, until equilibrium condition was achieved. This behavior is observed for each line in **Figures 2** and **3** throughout a gradual increase of the concentration ratio of (outlet/initial) concentrations of CO_2 (C/C_o).

The CO_2 adsorption was most intensive during 50 min. and thereafter remains unchanged until saturation was attained. Adsorption process to carbon dioxide for different temperatures (30, 50, 70, and 90°C) on AC reaches equilibrium for increased temperature. The same behavior is shown in **Figure 3** when using a high concentration of carbon dioxide but fast (30 min. to reach the equilibrium), due to high CO_2 concentration, and this behavior is compatible with previous results when using AC prepared from coconut residue to remove carbon dioxide [36]. It is noted that breakthrough curves become shorter and steeper for high temperatures; this indicates the adsorption process here is exothermic and that's compatible with previous results for some of the previous adsorption of carbon dioxide on the zeolite [1]. The adsorption of carbon dioxide carbon process was not affected by the presence of nitrogen gas, and this is due to the strength of the links formed by carbon dioxide with AC particles [36].

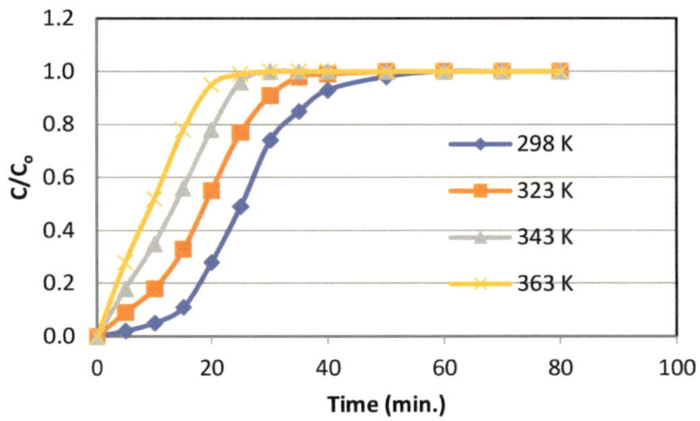

Figure 2.
Breakthrough curve for CO_2 adsorption onto AC (initial conc. = 10%vol., avg. particle diameter = 5 mm, and volumetric rate = 12 l/min).

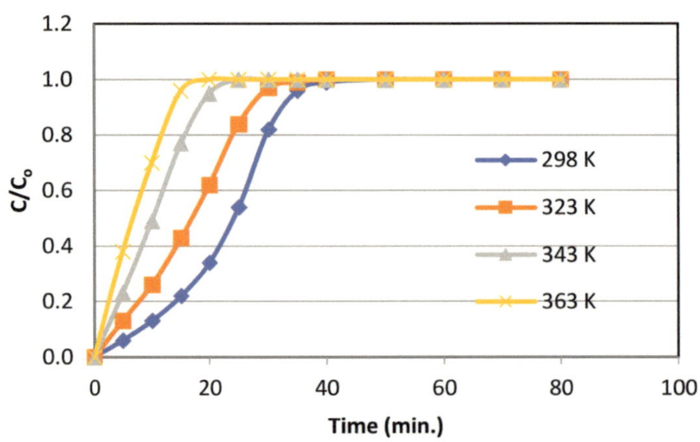

Figure 3.
Breakthrough curve for CO_2 adsorption onto AC (initial conc. = 13.725%vol., avg. particle diameter = 5 mm, and volumetric rate = 12 l/min).

In post-combustion process, the flue gas temperature is typically within the range of 50–120°C [36, 37]. Thus, an adsorption study was conducted at 30–90°C to investigate the CO_2 adsorptive properties at elevated temperatures. **Figures 4** and **5** show that the CO_2 adsorption capacity of solid adsorbent decreases with temperature, and it implies the existence of physical adsorption (physisorption) between the CO_2 molecules and carbonaceous adsorbent. Adsorption capacity decreased with increasing temperature because of exothermic adsorption process as shown in **Figures 4** and **5**. This behavior is also identical with the results of previous studies [3, 38]. The adsorption capacities recorded in **Figure 4** are 109.529, 74.57, 50.61, and 35.46 $mg_{(CO2)}/g_{AC}$, whereas they recorded in **Figure 5** as 129.651, 89.2, 53.079, and 35.546 $mg_{(CO2)}/g_{AC}$ at temperatures 30, 50, 70, and 90°C, respectively. Thus, the optimum temperature for the removal process is 30°C. It also notices that the adsorption process occurs in the beginning quickly and be a decline in the curves and clear because of the abundance of the active sites and the presence of small surface resistance on the surface of adsorbents, then more smoothness gradually less steep and alignment over time because of the fullness of all the active sites on the surface of adsorbents and that the process has become controlled by internal diffusion within the adsorbents in accordance with what has been presented previously [39]. Also the results of comparison for both **Figures 4** and **5** together note

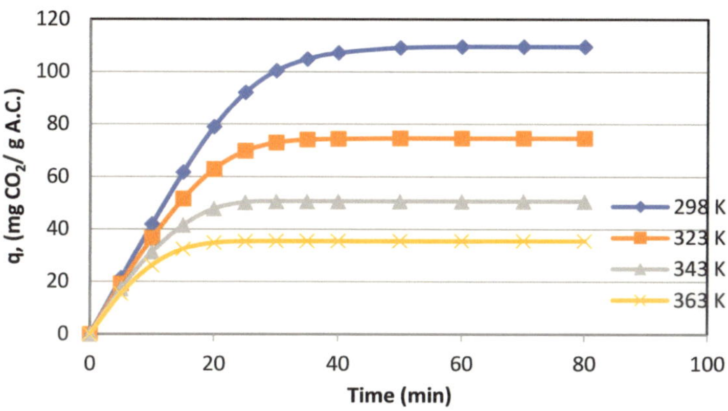

Figure 4.
Adsorption capacity of CO_2 onto AC (initial concn. = 10%vol., avg. particle diameter = 5 mm, and volumetric rate = 12 l/min.).

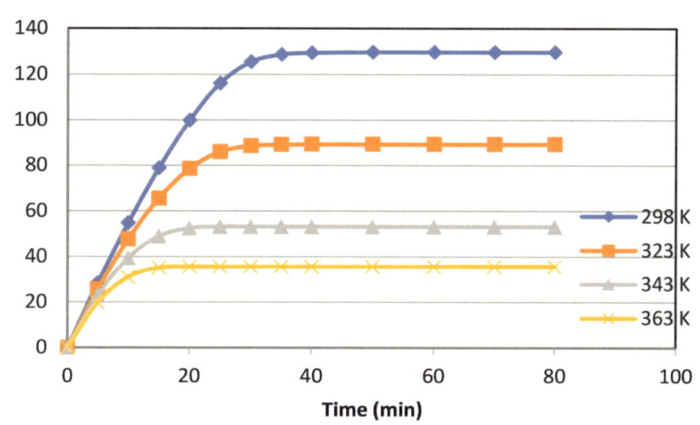

Figure 5.
Adsorption capacity of CO_2 onto AC (initial conc. = 13.725%vol., avg. particle diameter = 5 mm, and volumetric rate = 12 l/min.).

that the amount of the CO_2 adsorbed onto AC increases due to increase of the concentration difference of CO_2 between bulk and surface of AC leading to an increase of mass transfer [9, 40].

3.3 Equilibrium isotherm studies

The equilibrium data can be approximated using common and practical adsorption isotherms, which provide the basis for the design of adsorption systems. The amount of adsorbed CO_2 onto adsorbent (AC) as a function of its concentration at constant temperature can be described by different adsorption isotherm models (Langmuir, Freundlich, Temkin, and Dubinin-Radushkevich). The predicted isotherm constants for the CO_2 adsorption and the determination coefficient r^2 value from the linear regression method are shown in **Table 2**.

Based on tabulated data, the maximum capacity (Q_o) of AC and a_L values (Langmuir parameters) for CO_2 adsorption decreased with increasing temperature; this reveals a physisorption process occurred. The decline in values of maximum adsorption capacity with increased in the adsorption temperature is due to the exothermic nature of the CO_2 adsorption on AC. It is confirmed by n values higher than 1 in the Freundlich isotherm model that the adsorption is favorable for AC.

C_o	T	*Langmuir constants*		*Freundlich constants*			Q_{exp}	Q_L	r^2	Q_F	r^2
Vol. %	°C	a_L (l/mg)	Q_o (mg/g)	K_F (mg/g)		n (−)	(mg/g)	(mg/g)	—	(mg/g)	—
10	30	0.0492	120.482	0.00081		0.3865	109.53	108.03	0.997	522.49	0.955
	50	0.01835	99.009	0.04223		0.527	74.57	90.633	0.9976	685.26	0.954
	70	0.01076	81.301	0.2138		0.5998	50.61	50.92	0.9865	966.67	0.964
	90	0.00778	66.667	0.5181		0.6342	35.46	49.835	0.9829	1358.19	0.972
13.725	30	0.01171	161.29	0.002202		0.5324	129.65	129.693	0.9976	649.62	0.936
	50	0.0921	133.333	0.0933		0.5821	89.2	89.92	0.9919	1024.09	0.959
	70	0.0782	85.47	0.1711		0.5596	53.08	53.2934	0.9964	2453.47	0.982
	90	0.0466	67.11	0.6192		0.5996	35.55	35.78	0.9872	6014.59	0.984
C_o	**T**	**Temkin constants**		**D-R constants**			**Q_{exp}**	**Q_T**	**r^2**	**Q_{D-R}**	**r^2**
Vol. %	**°C**	**B (J/mol)**	**A (l/g)**	**ß (mol²/kJ²)**	**q_m (mg/g)**	**ε (kJ/mol)**	**(mg/g)**	**(mg/g)**	**—**	**(mg/g)**	**—**
10	30	22.305	0.746	0.0078	93.841	8.006	109.53	108.84	0.998	93.76	0.806
	50	22.582	0.167	0.0881	70.316	2.382	74.57	75.009	0.997	69.48	0.909
	70	19.114	0.092	0.221	51.07	1.506	50.61	50.965	0.994	49.16	0.955
	90	15.473	0.067	0.339	37.68	1.123	35.46	35.58	0.996	35.02	0.982
13.725	30	35.826	0.164	0.104	117.47	2193	129.65	131.45	0.989	116.78	0.86
	50	30.949	0.0813	0.308	86.59	1.27	89.2	89.93	0.996	84.62	0.923
	70	20.395	0.0642	0.519	54.806	0.981	53.08	53.24	0.998	52.21	0.968
	90	16.048	0.0378	1.276	39.186	0.626	35.55	35.63	0.9947	35.43	0.996

Table 2.
Parameters of isotherm models at different temperatures via linearized technique for adsorption of CO_2 onto AC.

In addition, the Dubinin-Radushkevich isotherm will provide a useful information related to the energy parameters, in terms of E (mean free energy of adsorption). The calculated E values which are within the range of 1.213–8 and 0.626–2.193 kJ/mol for both initial concentrations 10 and 13.725% vol., respectively, suggest that the CO_2 adsorption is physical in nature, as the magnitude of E is below 8 kJ/mol, whereas the value of $8 < E < 16$ is an indicator of the chemical adsorption [41]. Also, B (value of heat adsorption) in Temkin isotherm ranged between 15.47 and 35.826 J/mol indicating a physisorption process occurs.

On the basis of corresponding r^2 values and adsorption capacity estimated by each model shown in **Table 2**, the Langmuir model gives the best fit toward the experimental data over the entire temperature range. Therefore, it implies that the surface of the activated carbon is heterogeneous and a restricted monolayer CO_2 adsorption occurs, as results of adsorption CO_2 onto activated carbon prepared from coconut fiber studied by Hauchhum and Mahanta [3].

3.4 Kinetic studies

An analysis of kinetic adsorption process is a useful tool to estimate the time of residence for the adsorption process to complete and to determine the dynamics of adsorption and its performance in industrial scale of fixed bed or in flow-through systems. Thus, simple first-order, pseudo-first-order, pseudo-second-order, and intra-particle diffusion models were performed in this study. Kinetic parameters of these models are shown in **Table 3**.

Table 3 shows the simple first-order kinetic model for activated carbon did not fit well with the experimental data, with r^2 value found to be within the range of 0.4344–0.7873 and 0.319–0.734 for both initial concentrations 10 and 13.725% vol., respectively. Also, **Table 3** shows the pseudo-first-order kinetic model for activated carbon did not fit well with the experimental data, with r^2 value found to be within the range of 0.9521–0.967 also from 0.9175 to 0.9487 for both initial concentrations 10 and 13.725% vol., respectively.

Comparing the values of determination coefficients as stated in **Table 3**, pseudo-second-order model gives better fit than the pseudo-first-order and intra-particle diffusion models with experimental data, with r^2 value within the range of 0.963–0.996 and 0.955–0.998 for both initial concentrations 10 and 13.725% vol., respectively. Also, the values of adsorption capacity of equilibrium (q_e) were observed to decrease with respect to temperature. The kinetic energy of CO_2 adsorbed at elevated temperatures is high, and it leads to its increasing tendency to escape from the AC surface. Maroto-Valer et al. [42] reported that physisorption process involves high surface adsorption energy and molecule diffusion at elevated temperatures, which result in instability of the adsorbed gas on the surface of activated carbon, and consequently desorption process will occur.

In similarity to pseudo-first-order and pseudo-second-order models, the intra-particle diffusion model provides insight of the mechanism in adsorption process. Adsorption contains of few steps involved in the transfer of adsorbate (CO_2) from the phase of bulk to the solid surface of AC and is followed by the molecule diffusion into the interior of the pores of AC. Intra-particle diffusion is typically described as a slow process and is a limiting step in many adsorption processes. Theoretically, if the adsorption process obeys the intra-particle diffusion model, a straight linear plot that passes through the origin is expected. However, results of the variation of gradient with respect to time show that the intra-particle diffusion is not the sole rate-limiting step in this adsorption process. Note that the first

Kinetic model	C_o (vol.%)	Parameter	Temperature (°C)			
			30	50	70	90
Simple first-order	10	k_1	2.633	1.974	1.394	0.955
		C_o	4.1×10^5	1.75×10^{21}	1.67×10^{33}	6.73×10^{40}
		r^2	0.787	0.667	0.539	0.434
	13.725	k_1	3.293	2.404	1.457	1.091
		C_o	3.04×10^{17}	8.414×10^{36}	6.52×10^{56}	1.023×10^{80}
		r^2	0.734	0.624	0.427	0.319
Pseudo-first-order	10	k_1	0.0972	0.1521	0.1864	0.2367
		q_e	159.48	143.481	84.94	62.22
		r^2	0.967	0.96	0.915	0.952
	13.725	k_1	0.1508	0.182	0.2116	0.282
		q_e	292.89	195.45	78.25	49.47
		r^2	0.918	0.927	0.944	0.949
Pseudo-second-order	10	k_2	1.6×10^5	6.59×10^4	3.14×10^4	1.69×10^4
		q_e	144.93	87.719	55.56	37.45
		r^2	0.968	0.976	0.989	0.995
	13.725	k_2	2.88×10^5	1.29×10^5	5.76×10^4	3.23×10^4
		q_e	161.29	102.04	56.179	36.496
		r^2	0.96	0.982	0.955	0.998
Intra-particle diffusion	10	k_p	13.974	8.755	5.436	3.489
		c	7.686	12.25	13.025	11.775
		r^2	0.874	0.8123	0.7417	0.6811
	13.725	k_p	15.976	10.169	5.219	3.151
		c	14.137	17.633	17.693	14.578
		r^2	0.841	0.788	0.677	0.601

Table 3.
Kinetic parameters of CO_2 onto AC.

steeper region (2–4 min.$^{1/2}$) could be due to surface sorption, while the second region (4–9 min.$^{1/2}$) may be attributed by the intra-particle diffusion rate controlled.

3.5 Thermodynamic studies

The values of thermodynamic parameters of CO_2 adsorption process on AC based on Van't Hoff plot for Eqs. (22) and (23) are shown in **Figure 6**.

The estimated values of the thermodynamic parameters are tabulated in **Table 4**. For significant adsorption to occur, the Gibbs free energy change of adsorption (ΔG^o) must be negative [43]. **Table 4** shows that the (ΔG^o) was negative values for all five temperatures studied, which indicates the feasibility and spontaneity of the adsorption process. In addition, decreased negative ΔG^o value with increasing temperature implies that the CO_2 adsorption process is more favorable at 30°C rather than at 90°C; this behavior is also noticed by Rashidi et al. [44] and Hauchhum and Mahanta [3].

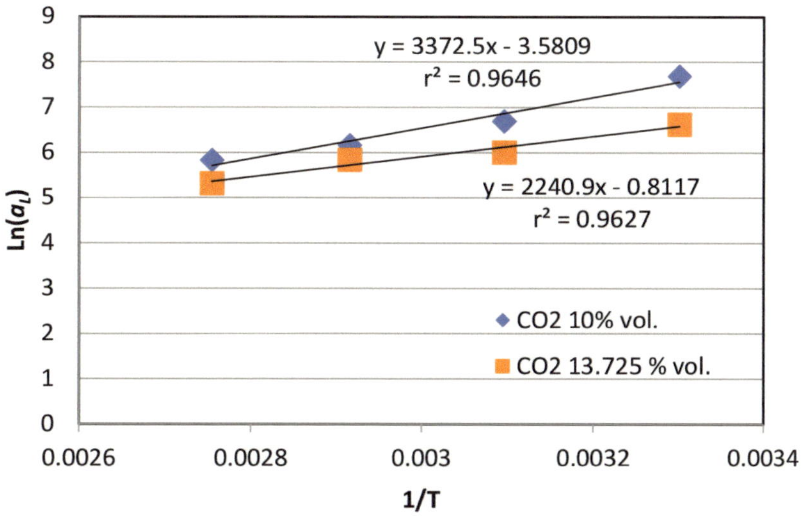

Figure 6.
Van't Hoff plot for adsorption of CO_2/AC system.

Conc.	Temp.	a_L	ΔG^o	ΔH^o	ΔS^o
(vol.%)	(K)	(mol/l)	(J/mol)	(J/mol)	(J/mol.K)
10	303	2164.79	−19347.21	−28038.965	−29.7716
	323	807.71	−17976.75		
	343	473.45	−17566.61		
	363	342.50	−17613.77		
13.725	303	753.38	−16688.24	−18630.843	−6.748
	323	405.31	−16124.99		
	343	344.14	−16065.94		
	363	204.91	−16063.38		

Table 4.
Thermodynamic parameters of CO_2 adsorption onto AC.

According to findings of the experimental data, the negative sign of ΔH^o value indicates an exothermic nature of the CO_2 adsorption process onto AC, and the negative value of ΔS^o suggests high orderliness of the adsorbate molecules (CO_2) upon adsorption. Zhao et al. [45] mentioned that the negative value of ΔS^o can be interpreted by the behavior of the CO_2 molecules upon the adsorption process, which is from randomized to an ordered form on the surface of the adsorbent. The decline in the entropy value upon the adsorption process is due to a lesser degree of freedom of the CO_2 molecules, due to minimum free space on the surface of AC. Moreover, the value of ΔH^o indicates the type of CO_2 adsorption process, whether it belongs to the physisorption or chemisorption. It has been reported that the value of ΔH^o for the physisorption process is <20 kJ/mol, while for the chemisorption process, the value is within 80–200 kJ/mol [45, 46]. Therefore, the calculated values of ΔH^o approximately ranging between 18 and 28 kJ/mol suggest that the CO_2 adsorption can be attributed to a physi-intra-particle diffusion adsorption process rather than a pure physisorption or chemisorption process. Also, this supports the isotherm study results that reveal the adsorption mechanism is physisorption and obeys Langmuir isotherm model.

3.6 Effects of interaction between gases in mixture

The adsorption amount for each component in a complex mixture of (CO_2, NO, and N_2) was compared with that under the single-component conditions, with the results shown in **Figure 7**. In the single-component condition, the adsorption amount was 109.24 mg_{CO2}/g_{AC} and 0.245 mg_{NO}/g_{AC} in 50 and 40 min for each one, respectively. When all of the components were present in a mixture of CO_2, NO, and N_2, the CO_2 adsorption amount decreased by 6%, and the NO adsorption amount also decreased by 7.6%. The adsorption capacity of CO_2 in complex mixture is not changed compared with that under the single-component conditions but is favorable due to decreased equilibrium time required. The adsorption capacities of complex mixture were 103.2 mg_{CO2}/g_{AC} in 40 min and 0.229 mg_{NO}/g_{AC} in 30 min. This decrease in adsorption capacity is due to the competition between both gases on the active sites, whereas the CO_2 adsorption capacity is higher than NO gas because of the presence of CO_2 gas at high concentrations. NO and CO_2 display fast breakthrough, and high adsorption amounts were observed in the pure component adsorption experiments. When the interaction effect of CO_2 and NO was considered, a very interesting phenomenon appeared. After the initial breakthrough, the CO_2 concentration descends to a minimum and then gradually ascends with the breakthrough ending point of NO. This is observed from the arrival time of the

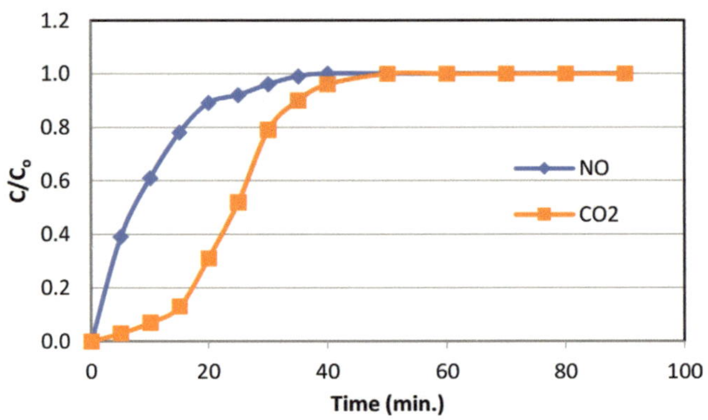

Figure 7.
Breakthrough curves for mixture gas (CO_2, NO, and N_2) (initial conc. = 10% vol. of CO_2, and 550 ppm of NO, avg. particle diameter = 5 mm, temperature = 30°C, and volumetric rate = 12 l/min.)

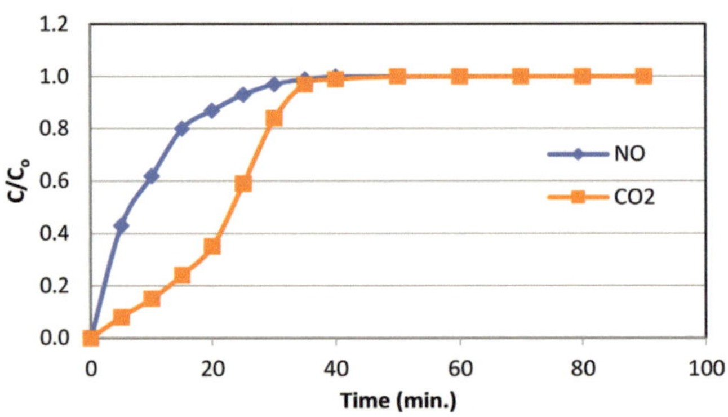

Figure 8.
Breakthrough curves for mixture gas (CO_2, NO, and N_2) (initial conc. = 13.725%vol. of CO_2, and 550 ppm of NO, avg. particle diameter = 5 mm, temperature = 30°C, and volumetric rate = 12 l/min.)

equilibrium state which was 30 min for the NO gas and 40 min for the CO_2 gas; the same behavior is shown for the single-component conditions; the difference between them was reached in the equilibrium stage fast approximately 10 min. In **Figure 8** the adsorption capacities of complex mixture were 124.4 mg_{CO2}/g_{AC} in 40 min and 0.22 mg_{NO}/g_{AC} in 30 min by using different initial concentrations of gases. This is confirmed by the fact that free Gibbs energy values of both gases are very different, as they have NO gas higher in value, while in the case of CO_2, they are much lower. This behavior is shown in previous study for adsorption on activated carbon prepared from coconut husk residues [47].

4. Conclusion

In this study, the fixed bed adsorption of carbon dioxide from CO_2/N_2 mixtures on activated carbon was studied. The adsorption dynamics was investigated at different operating temperatures (30–90°C). The results show that the low-cost activated carbon can be prepared from olive trees as potential carbonaceous material serving as porous media for CO_2 capture. Based on the experimental results, it is concluded that the CO_2 adsorption onto the olive tree activated carbon follows the physisorption behavior, whereby the CO_2 adsorption capacity decreases with respect to increasing temperature. Based on the equilibrium models of isotherm used herein to fit the experimental data of adsorption, the Langmuir model was the best fit with experimental data over the whole temperature range, due to the highest estimated adsorption capacity and determination coefficient (r^2) closeness to unity, thus implying a perfect fit to the experimental data. Besides, thermodynamic analysis proves that the CO_2 adsorption is a spontaneous process at low temperature, physisorption, and intra-particle diffusion and exothermic in nature. Also, the negative values of the entropy of the adsorption manifest the restricted randomness of the adsorbate molecules on the surfaces of adsorbent. CO_2 adsorption capacity has been reduced slightly when NO appears, but the process of adsorption has been faster as a result of competition on carbon-active sites.

Olive trees are dominant and easily available in the Mediterranean countries generally and especially Libya, and because charcoal is prepared from olive trees, it is cheap in Libya. According to the obtained result, this study concludes that AC prepared from olive trees can be considered as adequate for designing a fixed bed cycle to separate carbon dioxide from flue gases and serve as a benchmark while searching for inexpensive and superior activated carbon production in future studies that concerned of capturing CO_2 from flue gases of the industrial sectors (such as cement plants and power stations) that are prevailing in Libya.

Author details

Hesham G. Ibrahim[1*] and Mohamed A. Al-Meshragi[2]

1 Mechanical Engineering Department, Faculty of Marine Resources, Al-Asmarya Islamic University, Zliten, Libya

2 Chemical Engineering Department, University of Tripoli, Tripoli, Libya

*Address all correspondence to: h.ibrahim@asmarya.edu.ly

References

[1] Dantas TLP, Luna FMT, Silva Jr IJ, Torres AEB, de Azevedo DCS, Rodrigues AE, et al. Modeling of the fixed-bed adsorption of carbon dioxide and a carbon dioxide-nitrogen mixture on zeolite 13x. Brazilian Journal of Chemical Engineering. 2011;**28**(3): 533-544. DOI: 10.1590/ S0104-66322011000300018

[2] Zhao Z, Cui X, Ma J, Li R. Adsorption of carbon dioxide on alkaline-modified zeolite 13X adsorbents. The International Journal of Greenhouse Gas Control. 2007;**1**(3):355-357. DOI: 10.1016/s1750-5836(07)00072-2

[3] Hauchhum L, Mahanta P. Kinetic, thermodynamic and regeneration studies for CO_2 adsorption onto activated carbon. International Journal of Advanced Mechanical Engineering. 2004;**4**(1):27-32. Available from: https://pdfs.semanticscholar.org/eacc/3f82336a3276744eda9a898c4621da7d357d.pdf

[4] Grande CA, Rodrigous AE. Electric swing adsorption for CO_2 removal from flue gases. International Journal of Greenhouse Gas Control. 2008;**2**: 194-202. DOI: 10.1016/s1750-5836(07)00116-8

[5] Cavenati S, Grande CA, Rodrigues AE. Separation $CH_4/CO_2/N_2$ mixtures by layered pressure swing adsorption for upgrade of natural gases. Chemical Engineering Science. 2006;**61**: 3893-3906. DOI: 10.1016/j.ces.2006.01.023

[6] Chen LC, Peng PY, Lin LF, Yang TCK, Huang CM. Facile preparation of nitrogen-doped activated carbon for carbon dioxide adsorption. Aerosol and Air Quality Research. 2014;**14**:916-927. DOI: 10.4209/aaqr.2013.03. 0089

[7] Zeinali F, Ghoreyshi AA, Najafpour GD. Adsorption of dichloromethane from aqueous phase using granular activated carbon: Isotherm and breakthrough curve measurements. Middle-East Journal of Scientific Research. 2010;**5**(4):191-198. DOI: 10.1080/00986445.2011.584354

[8] Alshuiref AA, Ibrahim HG, Ben Mahmoud MM, Maraie AA. Treatment of wastewater contaminated with Cu(II) by adsorption onto acacia activated carbon. Journal of Marine Sciences and Environmental Technologies (JMSET). 2017;**3**(2):25-36 Available from: http://www.asmarya.edu.ly/journal2/wp-content/uploads/2018/04/JMSET03-3-2-2017.pdf

[9] Kuboňová L, Obalová L, Vlach O, Troppová I, Kalousek J. Modeling of NO adsorption in fixed bed on activated carbon. Chemical and Process Engineering. 2011;**32**(4):367-377. DOI: 10.2478/v10176-011-0029-z

[10] McCabe W, Smith J, Harriott P. Unit Operations of Chemical Engineering. 7th ed. NY, USA: McGraw Hill Chemical Engineering Series; 2004. ISBN-10: 0072848235

[11] Khalili S, Ghoreyshi A, Jahanshai M. Carbon dioxide captured by multi-walled carbon nanotube and activated charcoal: A comparative study. Chemical Industry & Chemical Engineering. 2013;**19**(1):153-164. DOI: 10.2298/ciceq120217050k

[12] Ibrahim HG. Removal and Recovery of Chromium from Aqueous Solutions. Saarbrucken, Germany: LAP Lambert Academic Publishing GmbH and Co. KG; 2010; ISBN-10: 3838339037

[13] Özacar M. Adsorption of phosphate from aqueous solution onto alunite. Chemosphere. 2003;**51**(4):321-327. DOI: 10.1016/s0045-6535(02)00847-0

[14] Choy KK, McKay G, Porter JF. Sorption of acid dyes from effluents using activated carbon. Resources,

Conservation and Recycling. 1999;**27**(1–2):57-71. DOI: 10.1016/s0921-3449(98)00085-8

[15] Carrasco-Marin F, Lopez-Ramon MV, Moreno-Castilla C. Applicability of the Dubinin-Radushkevich equation to carbon dioxide adsorption on activated carbons. Langmuir. 1993;**9**(11):2758-2760. DOI: 10.1021/-la00035a002

[16] Lin SH, Juang RS. Heavy metal removal from water by sorption using surfactant-modified montmorillonite. Journal of Hazardous Materials. 2002;**92**(3):315-326. DOI: 10.1016/s0304-3894(02)00026-2

[17] Fadali OA, Magdy YH, Daifullah AAM, Ebrahiem EE, Nassar MM. Removal of chromium from tannery effluents by adsorption. Journal of Environmental Science and Health. 2004;**39**(2):465-472. DOI: 10.1081/ese-120027537

[18] Yang H, Xu Z, Fan M, Gupta R, Slimane RB, Bland AE, et al. Progress in carbon dioxide separation and capture: A review. Journal of Environmental Sciences. 2008;**20**:14-27. DOI: 10.1016/S1001-0742(08)60002-9

[19] Raven KP, Jain A, Loeppert RH. Arsenite and arsenate adsorption on ferrihydrite: Kinetics, equilibrium, and adsorption envelopes. Environmental Science & Technology. 1998;**32**(3):344-349. DOI: 10.1021/es970421p

[20] Al-Meshragi M, Ibrahim HG, Okasha AY. Removal of trivalent chromium from aquatic environment by cement kiln dust: Batch studies. AIP Conf. Proc. 2009;**1127**(1):74-85. DOI: 10.1063/1. 3146200

[21] Hossain MA, Kumita M, Michigami Y, Mori S. Optimization of parameters for Cr (VI) adsorption on used black tea leaves. Adsorption. 2005;**11**(5–6):561-568. DOI: 10.1007/s10450-005-5613-4

[22] Eligwe CA, Okolue NB. Adsorption of iron (II) by a Nigerian brown coal. Fuel. 1994;**73**(4):569-572. DOI: 10.1016/0016-2361(94)90042-6

[23] Sparks DL. Kinetics of Soil Chemical Processes. 1st ed. Massachusetts, USA: Academic Press; 1989. DOI: 10.1016/b978-012656446-4/50007-4

[24] Annadurai G, Juang RS, Lee DJ. Use of cellulose-based wastes for adsorption of dyes from aqueous solutions. Journal of Hazardous Materials. 2002;**92**(3):263-274. DOI: 10.1016/s0304-3894(02)00017-1

[25] Özacar M, Şengil IA. Adsorption of reactive dyes on calcined alunite from aqueous solutions. Journal of Hazardous Materials. 2003;**98**(1–3):211-224. DOI: 10.1016/s0304-3894(02)00358-8

[26] Ho YS, Chiang CC. Sorption studies of acid dye by mixed sorbents. Adsorption. 2001;**7**(2):139-147. DOI: 10.1023/A:1011652224816

[27] Chiou MS, Li HY. Equilibrium and kinetic modeling of adsorption of reactive dye on cross-linked chitosan beads. Journal of Hazardous Materials. 2002;**93**(2):233-248. DOI: 10.1016/s0304-3894(02)00030-4

[28] Wu FC, Tseng RL, Juang RS. Kinetic modeling of liquid-phase adsorption of reactive dyes and metal ions on chitosan. Water Research. 2001;**35**(3):613-618. DOI: 10.1016/s0043-1354(00)00307-9

[29] Wu FC, Tseng RL, Juang RS. Kinetics of color removal by adsorption from water using activated clay. Environmental Technology. 2001;**22**(6):721-729. DOI: 10.1080/09593332208618235

[30] Ho YS, McKay G. Pseudo-second order model for sorption processes.

Process Biochemistry. 1999;**34**(5): 451-465. DOI: 10.1016/s0032-9592(98) 00112-5

[31] Srivastava SK, Tyagi R, Pant N. Adsorption of heavy metal ions on carbonaceous material developed from the waste slurry generated in local fertilizer plants. Water Research. 1980; **23**(9):1161-1165. DOI: 10.1016/ 0043-1354(89)90160-7

[32] Shiue A, Hu SC, Chang SM, Ko TY, Hsieh A, Chan A. Adsorption kinetics and breakthrough of carbon dioxide for the chemical modified activated carbon filter used in the building. Sustainability. 2017;**9**(9):1-13. DOI: 10.3390/su9091533

[33] Demirbaş E, Kobya M, Öncel S, Şencan S. Removal of Ni(II) from aqueous solution by adsorption onto hazelnut shell activated carbon: Equilibrium studies. Bioresource Technology. 2002;**84**(3):291-293. DOI: 10.1016/s0960-8524(02)00052-4

[34] Sahmoune MN, Louhab K, Boukhiar A, Addad J, Barr S. Kinetic and equilibrium models for the biosorption of Cr(III) on Streptomyces rimosus. Toxicological and Environmental Chemistry. 2009;**91**(7):1291-1303. DOI: 10.1080/02772240802613731

[35] Wang S, Li H. Dye adsorption on unburned carbon: Kinetics and equilibrium. Journal of Hazardous Materials. 2005;**126**(1–3):71-77. DOI: 10.1016/j.jhazmat.2005.05.049

[36] Rashidi NA, Yusup S, Loong LH. Kinetic studies on carbon dioxide capture using activated carbon. Chemical Engineering Transaction, AIDIC Publications. 2013;**35**:361-366. DOI: 10.3303/CET1335060

[37] Kaithwas A, Prasad M, Kulshreshtha A, Verma S. Industrial wastes derived solid adsorbents for CO_2 capture: A mini review. Chemical Engineering Research and Design. 2012;**90**:1632-1641. DOI: 10.1016/j.cherd.2012.02.011

[38] Guo B, Chang L, Xiel K. Adsorption of carbon dioxide on activated carbon. Journal of Natural Gas Chemistry. 2006; **15**(3):223-229. DOI: 10.1016/S1003-9953 (06)60030-3

[39] Li A, Wu H, Zhang Q, Zhang G, Long C, Fei Z, et al. Thermodynamic study of adsorption of phenolic compounds from aqueous solution by a water-compatible hypercrosslinked polymeric adsorbent. Chinese Journal of Polymer Science. 2004;**22**(3):259-267

[40] Thomas WJ, Crittenden B. The Literature of Adsorption. Adsorption Technology and Design. Massachusetts, USA: Butterworth-Heinemann; 1989. DOI: 10.1016/b978-075061959-2/ 50009-4

[41] Soltani DC, Safari M, Rezaee A, Godini H. Application of a compound containing silica for removing ammonium in aqueous media. Environmental Progress & Sustainable Energy. 2014;**34**(1):105-111. DOI: 10.1002/ep.11969

[42] Maroto-Valer MM, Tang Z, Zhang Y. CO_2 capture by activated and impregnated anthracites. Fuel Processing Technology. 2005;**86**: 1487-1502. DOI: 10.1016/j. fuproc.2005.01.003

[43] Saha P, Chowdhury S. Insight into adsorption thermodynamics. In: Tadashi M, editor. Thermodynamics. Rijeka: InTech; 2011. DOI: 10.5772/13474

[44] Rashidi NA, Yusup S, Borhan A. Isotherm and thermodynamic analysis of carbon dioxide on activated carbon. Procedia Engineering. 2016;**148**: 630-637. DOI: 10.1016/j. proeng.2016.06.527

[45] Zhao Y, Wang D, Xie H, Won SW, Cui L, Wu G. Adsorption of Ag (I) from aqueous solution by waste yeast: Kinetic, equilibrium and mechanism studies. Bioprocess and Biosystems Engineering. 2015;**38**(1):69-77. DOI: 10.1007/s00449-014-1244-z

[46] Liang S, Guo X, Feng N, Tian Q. Isotherms, kinetics and thermodynamic studies of adsorption of Cu^{2+} from aqueous solutions by Mg^{2+}/K^{+} type orange peel adsorbents. Journal of Hazardous Materials. 2010;**174**(1–3): 756-762. DOI: 10.1016/j.jhazmat.2009.09.116

[47] Yi H, Wang Z, Liu H, Tang X, Ma D, Zhao S, et al. Adsorption of SO_2, NO, and CO_2 on activated carbons: Equilibrium and thermodynamics. Journal of Chemical & Engineering Data. 2014;**59**(5):1556-1563. DOI: 10.1021/je4011135

Chapter 8

Carbon Capture and Storage (CCS): Geological Sequestration of CO_2

Nediljka Gaurina-Međimurec and Karolina Novak Mavar

Abstract

The European Union greenhouse gas emission reduction target can be achieved only by applying efficient technologies, which give reliable results in a very short time. Carbon capture and storage (CCS) into geological formations covers capturing CO_2 at the large point sources, its transportation and underground deposition. The CCS technology is applicable to different industries (natural gas processing, power generation, iron and steel production, cement manufacturing, etc.). Due to huge storage capacity and existing infrastructure, depleted hydrocarbon reservoirs are one of the most favourable storage options. In order to give overall cross section through CCS technology, implementation status and other relevant issues, the chapter covers EU regulation, technology overview, large-scale and pilot CCS projects, CO_2-enhanced oil recovery (EOR) projects, geological storage components, CO_2 storage capacity, potential CO_2 migration paths, risk assessment and CO_2 injection monitoring. Permanent geological sequestration depends on both natural and technical site performance. Site selection, designing, construction and management must ensure acceptable risk rates of less than 1% over thousands of years.

Keywords: carbon capture and storage (CCS), geological sequestration, enhanced oil recovery, trapping mechanisms, risk assessment, monitoring

1. Introduction

Global warming issue and commitments towards reducing greenhouse gas emissions of at least 40% in 2030 and up to 95% in 2050 compared to 1990 level have initiated the development of certain strategies for CO_2 removal from the atmosphere, which recognised storage in underground formations as a most practical and suitable option. Although potential underground formation could be in the form of depleted oil and gas fields, deep saline formations or deep unmineable coal seams, commercial implementation is only possible if acceptable risk level is ensured. Huge practice, existing infrastructure and remaining storage capacities are the most important advantages of using depleted hydrocarbon reservoirs for those purposes. Furthermore, residual oil production, when carbon capture and storage is connected to enhanced oil recovery, is additional initiative. On the other hand, lack of research when it comes to other storage options requires different research programs to be performed in order to confirm projects feasibility and the safety of technology.

Formation storage possibility has to be defined through characterisation and assessment of potential storage complex, comprising data collection, static and dynamic modelling, sensitivity characterisation and risk assessment. Underground storage must meet relevant capacity and injectivity requirements, while storage efficiency depends on different physical and geochemical trapping mechanisms, which occur during the storage lifecycle [1]. Nevertheless, permanent storage is ensured by existing geological and equipment barriers; a certain risk of CO_2 migration has to be considered, assessed and controlled [2]. Special attention must be paid to the injected fluid migration issue, which implies identification of potential migration routes, such as faults and fractures, wells (active and abandoned) and seal rocks [3, 4]. In line with legal requirements, performed risk analysis and established monitoring plan, the effectiveness of storage complex has to be constantly evaluated. Comprehensive monitoring, which covers CO_2 plume tracking and surrounding environment monitoring, represents a very important part of the overall risk management strategy.

2. CCS deployment legal background

The international climate goal, set within the United Nations Framework Convention on Climate Change (UNFCCC) in Paris in 2015, seeks the limitation of the average temperature increase to below 2°C, compared to preindustrialisation reference level. That quite ambitious climate target depends on economy decarbonisation through increasing energy efficiency, enhancing the share of renewables in energy production and reducing greenhouse gas emissions. In order to achieve low carbon economy, the EU strategy targeted greenhouse gases emission reduction by 40% by 2030, and up to 95% by 2050 compared to the base year (1990) level [5, 6].

However, despite the efforts to enhance "green energy" sources, the society is still largely dependent on fossil fuels and it is evident that conventional carbon technologies cannot be removed easily from the industry processes in close future. Therefore, a systematic approach is needed.

The EU Directive 2009/31/EC on the geological storage of carbon dioxide [7] entered into force in 2009, establishing a legal framework for safe CO_2 geological sequestration in Europe. The Directive attempted to prevent any significant CO_2 leakage risk or damage to health and/or the environment by setting requirements for the entire storage cycle. It excludes potable water aquifers and tectonically active zones as potential sites for permanent disposal of CO_2.

The EU-requested emission reductions are expected to be achieved through the main instrument—the European Emission Trading Scheme (EU ETS) (**Figure 1**). The system is based on the EU Directive 2003/87/EC, establishing a scheme for greenhouse gas emission allowance trading within the Community [8]. It operates on a cap and trade principle, which considers behaviour in line with installations emission permits and market trading of EU emission unit allowances. Temporarily, the third phase of the system is operational (2013–2021). The main issue at the beginning of the third trading period was the imbalance between allowances supply and demand on the market, caused mainly by lower industrial activity. In order to overcome such unsustainable situation and increase the CO_2 price, which would encourage system participants to apply emission reduction measures comprising the CCS projects, a radical legislation revision was needed. It included the increase in the allowances reduction factor, auctioning the postponement of 900 million of allowances and establishment of the market stability reserve [9].

Carbon capture and storage technology is often observed as a transitional solution to low-carbon economy, due to possibility of further usage of fossil fuels in power

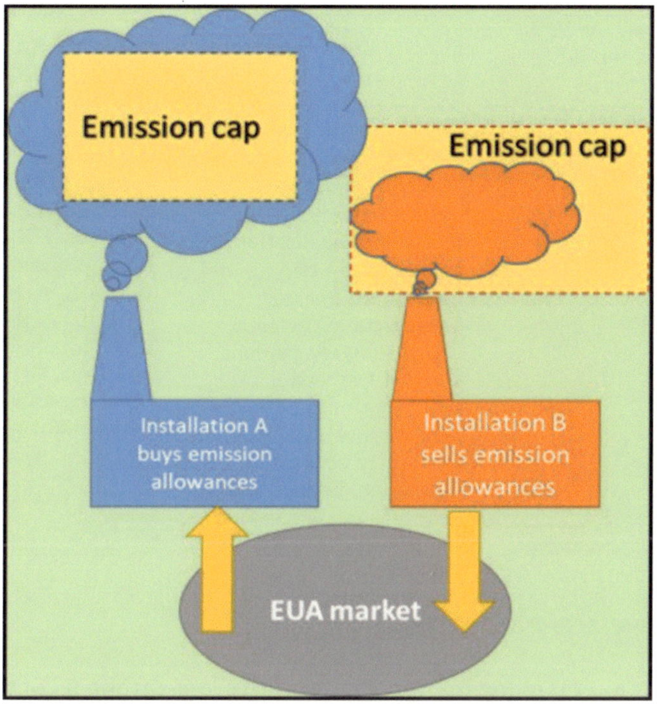

Figure 1.
The European Emission Trading Scheme (EU ETS) principles.

generation while simultaneously reducing CO_2 emission [10]. Since the demanding climate goals require about 4000 Mt/y of CO_2 to be removed from the atmosphere by 2040 [11], a lot of further effort has to be invested. Inclusion of CCS in clean development mechanisms (CDMs) is one step ahead in its global deployment [12].

The success of the CCS project is only possible if stable, clear and efficient regulatory framework and supporting public acceptance are ensured [13]. A political decision on CCS is influenced by different factors, such as national CO_2 emission level and emission reduction commitments, available storage capacity and public awareness. This means that most of the research and development activities occur in the states with the highest emissions intensity (e.g., Germany, the UK, Italy, France, Spain, the Netherlands and Norway). On the other hand, strong local public resistance (e.g., in Denmark, Germany, the UK, Poland and the Netherlands) resulted with the cancellation of more projects and the postponement of CO_2 storage acceptance [14].

Still, most of the EU Member States transposed the Directive without any restrictions and continue to support research in order to improve the technology (**Figure 2**).

Since the CCS initiatives in the EU originate from climate changes mitigation intention, projects in North America are mostly connected to the EOR activities, with CO_2 sales as a major incentive. Viability of such projects is strongly dependent on the oil price.

Due to instability of market oil prices, financial support is crucial to provide a certain level of certainty. CCS projects are supported by different policies at Federal, State and local levels. The Department of Energy (DOE) provides financial assistance and grants in line with the Energy Improvement and Extension Act (2008) and the American Recovery and Reinvestment Act (2009) [13]. In EU, additional funding may refer to the EU Energy Program for Recovery (EEPR), the NER300, FP7 or some national government funding schemes [15]. The ETS Innovation Fund is a new EU funding scheme, scheduled for 2021. Based on the

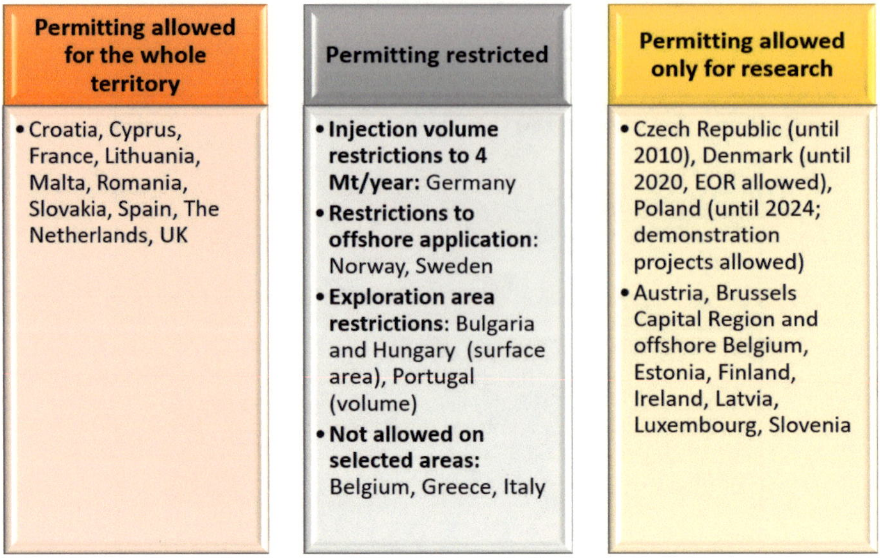

Figure 2.
CO_2 storage permitting in European countries [14].

NER300 platform, it is going to support innovative low-carbon technologies, including CCS demonstration projects, by monetizing 400 million of CO_2 emissions unit allowances (EUA) from the New Entrants' Reserve [16].

3. CCS technology overview

Capturing CO_2 from the exhaust gases generated by different energy intensive industries (e.g., power generation, oil refineries or iron, steel and cement production), its transportation and permanent sequestration are fundamental parts of the CCS processes.

Exhaust gas is a mixture, which, besides nitrogen, steam, particulate matters and some other pollutants, contains only a small share of CO_2 (3–15%). That means that pure CO_2 must be extracted using different capture technologies: (a) pre-combustion capture system, (b) post-combustion capture system, (c) oxyfuel combustion system and (d) industrial separation (**Figure 3**). Technology selection depends on the concentration of CO_2 in the gas stream, pressure and fuel type [1, 17].

A **pre-combustion** capture processes comprise adding steam or oxygen to primary fuel, which results in synthesis gas (gas containing H_2 and CO) production. Further reaction of CO and steam in the shift reactor produces a mixture of H_2 and CO_2 in concentration between 5 and 15% volume. After separation, CO_2 is extracted by physical or chemical adsorption. In a **post-combustion** capture system, CO_2 is extracted from nitrogen after combustion by different physical or chemical solvents, or it is separated by adsorbents or membranes. This common technology can be an upgrade to existing thermal power plants and different industrial facilities, etc. An **oxyfuel combustion** capture system considers oxygen addition in the process of fossil fuel combustion, resulting in more concentrated CO_2 stream (more than 80% volume), which is prone to easier separation. Although this technology is simple and highly efficient in CO_2 removal, wide application is still prevented by the high cost of pure oxygen production. **Industrial separation** has had the longest usage: the CO_2, as unwanted compound, is separated in different industrial processes, comprising natural gas, hydrogen and ammonia production [1, 2, 17].

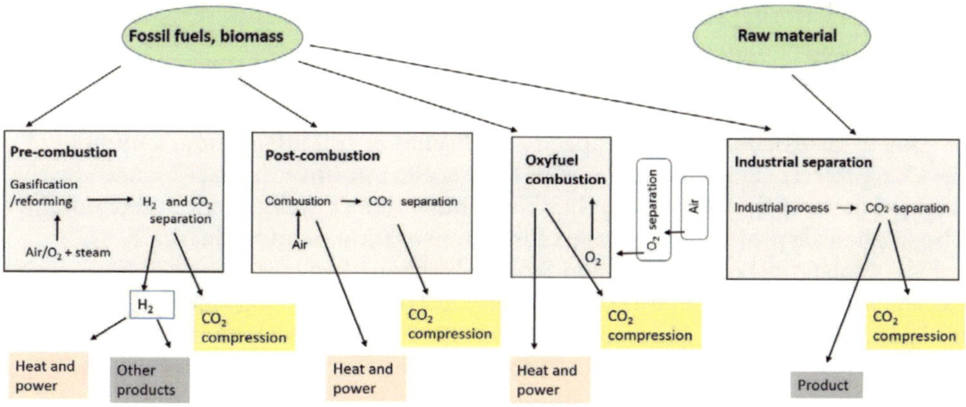

Figure 3.
Carbon capture processes [2].

The Carbon Capture R&D program has been implemented by the US National Energy Technology Laboratory (NETL) in order to develop cost-effective technologies based on different concepts (solvent, sorbent or membrane) [18].

After capturing, the CO_2 can be transported at solid, gaseous or liquid state or in the form of supercritical fluid. Although ships can be used, pipeline transport is often preferred as the most practical and the cheapest solution.

Application of CCS compared to other carbon sequestration options is preferred due to costs. The cost of geological storage of CO_2 depends on several factors such as the depth of the storage formation, the number of wells needed for injection and whether the project is onshore or offshore. For instance, capture system installed at fossil fuel power plant is between 15 and 75 USD/t (CO_2), where the coal-fired plants are the higher cost option. The costs are something lower in case of hydrogen, ammonia production or gas sweetening (from 5 to 55 USD/t (CO_2), while application to other industries is even more expensive, with costs between 25 and 115 USD/t (CO_2). Taking into consideration the costs of transportation of 5–40 Mt/y CO_2 by pipeline, which are on the level of 1–8 USD/t (CO_2), and geological storage and monitoring costs, which range from 0.6 to 8 USD/t (CO_2), it can be concluded that capture costs make up the majority of the price. However, considering the largest emissions belong to the fossil fuel power plants, it is important that research priority is focused on developing cost-effective capture technologies for power sector [19].

4. CCS projects

As is the case with all new technologies, implementation of CCS is facing different obstacles, which prevent a shift from the project planning phase to construction and operation phase. Commercial scale implementation requires a certain level of experience in technical, operational and economic feasibility of projects, which is substantial for risk decreasing and cost reduction.

Several decades of worldwide implementation of CCS research programs have resulted in a huge amount of experience and important knowledge on carbon capture and storage technology. The data obtained during large- and small-scale projects implementation are collected by different associations. Comprehensive databases founded by, for example, Carbon Capture and Sequestration Technologies at the Massachusetts Institute of Technology (MIT) [15], Global CCS Institute [20], National Energy Technology Laboratory (NETL) [18], Zero

Emissions Platform [21], British Geological Survey [22], etc., can serve as a valuable source of information in further research and design [2].

A large-scale facility captures at least 0.8 Mt of CO_2 from a coal-based facility for power generation or at least 0.4 Mt of CO_2 from other industry on yearly basis [20].

Due to insufficient capture capacity or absence of full integration, a number of the CCS projects cannot be declared as large scale, but since they are focused on the targeted parts of the CCS chain, they contribute to the development of technology. The small-scale projects can be used for demonstration or on a pilot scale.

The Global Carbon Capture and Storage Institute database counts 23 large-scale CCS facilities both in operation and under construction, having capture capacity of approximately 30 Mt/y. Realisation of further 5 projects, which are now in advanced planning phase, as well as another 15 projects, which are in early planning, could significantly increase capture capacity by more than 60 Mt/y.

Temporarily ongoing large-scale CCS projects are located in the USA, Canada, China, Saudi Arabia, United Arab Emirates and Europe. In Europe, the lack of national policy support and negative public opinion resulted in cancellation of some of the most promising CCS projects. However, successful operation of two Norwegian large-scale projects (Sleipner and Snøhvit) is enabled by high national carbon taxation. Future CCS activities in Europe are going to be expanded to two new offshore storage projects: Norway full chain CCS and Port of Rotterdam CCUS Backbone Initiative (Porthos).

Some of CCS projects are in the advanced planning or in the early planning phase. They are going to geologically store emissions from power generation and chemical industry. As regards CO_2 capture process, high cost of oxyfuel technology is the reason that only post-combustion technology has been applied [2, 20].

According to Carbon Capture and Sequestration Technologies at MIT database, there are substantial numbers of small-scale demonstration and pilot projects worldwide applied on different industries. Most of them are performed in Asia (China, Japan and South Korea), but also and to a lesser extent in the North America and Europe [15].

4.1 CO_2-EOR projects

Production from oil reservoirs is carried out in three phases: primary, secondary and tertiary. During the *primary recovery stage*, the reservoir pressure is sufficient to force the oil to the surface and recovery factor is typically 5–15%. During exploitation, reservoir pressure decreases and at one point, it becomes insufficient to force the oil to the surface. After that, *secondary recovery* methods are applied. They include water injection or natural gas reinjection to increase the reservoir pressure or gas lift (injection of gas into an active well to reduce the density of fluid in the well). The typical recovery factor from secondary operations is about 30%. Further increase of oil production is possible by the application of *tertiary oil recovery methods* or enhanced oil recovery (EOR) methods (including thermal recovery, chemical flooding and miscible gas injection), which increase the mobility of oil. Tertiary recovery provides additional production of 5–15% of oil.

CO_2-EOR is one of the tertiary oil recovery methods. The petroleum industry has been injecting CO_2 into partially depleted oil reservoirs for dozens of years. It is based on injection of CO_2 and usually water into the oil reservoir with the aim to enhance oil recovery by maintaining pressure in the reservoir and by improving oil ability to flow in the direction of the production well (**Figure 4**).

The CO_2 is produced along with the oil and then recovered and reinjected to recover more oil. When the maximum amount of oil is recovered from the reservoir, the CO_2 is then "sequestered" in the underground geologic zone that formerly contained oil and the well is shut in, permanently sequestering the CO_2.

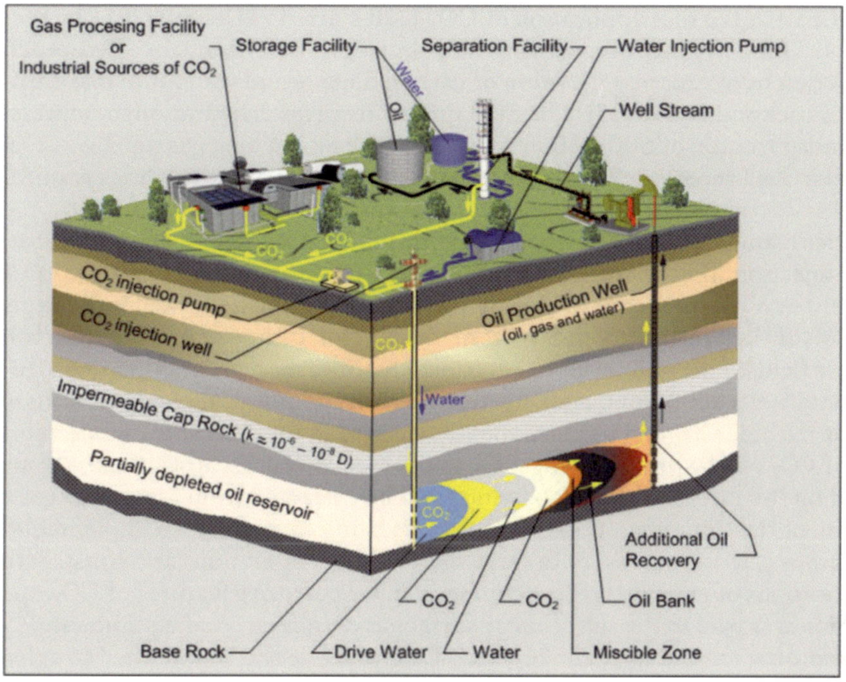

Figure 4.
The process of CO_2 and water injection in order to improve oil recovery [23].

EOR sites offer several advantages such as (1) well-understood geology and geologic seals, (2) proven capacity to hold volumes of CO_2 and (3) existing infrastructure such as surface facilities, pipelines, injection and monitoring wells.

CO_2-EOR can be employed onshore and offshore. It could lead to negative storage costs of 10–16 US$/t CO_2 for oil prices of 15–20 US$ per barrel and more for higher oil prices [1].

CO_2-EOR was first attempted in 1972 in Scurry County, Texas. In the 1970s, Shell was one of the first companies to inject naturally occurring carbon dioxide (CO_2) to increase oil recovery from fields in Texas, USA [24].

While initial CO_2-EOR developments used naturally occurring carbon dioxide deposits, technologies have been developed to inject CO_2 created as by-products from industrial operations. For example, Dakota Gasification Company's plant in Beulah, North Dakota, is producing CO_2 and delivering it by a 204-mile pipeline to the Weyburn oil field in Saskatchewan, Canada.

According to the CCS institute database, within the last 2 years, four large-scale projects were launched. Large-scale Emirates Steel Industries (ESI) CCS project running in Abu Dhabi represents the first application of CCS to iron and steel industry, where 0.8 Mt/y of CO_2 is injected underground for the purpose of hydrocarbon recovery [2, 20]. The Illinois Industrial CCS project enabled the capture of 1.0 Mt/y of CO_2 generated at the corn to ethanol facility in Decatur (Illinois, USA) and its permanent geological disposal, while the Petra New Carbon Capture project in Texas stands out for the largest power plant post-combustion CO_2 capture system. Captured gas at 1.4 Mt/y capacity is transported by pipeline and injected for EOR purposes. Another recent example where CO_2 is injected to improve oil recovery is the Chinese CNPC Jilin Oil Field CO_2 EOR project. After 12 years of testing, commercial operation started in 2018. The CO_2 source is at a natural gas processing plant. Capturing capacity is on the level of 600,000 t/y of CO_2.

In Croatia, the first application of CO_2-EOR started in October 2014 by the INA—Oil Industry Ltd. oil company. The project's aim is to enhance hydrocarbon production by alternating injection of carbon dioxide and water into mature oil fields Žutica and Ivanić [25]. The *EOR* project involves dehydration, compression and transportation of 600,000 m^3/day of CO_2 by 88 km long gas pipeline (20 in.) from the Gas Processing Facilities Molve to the Fractionation Facilities Ivanić Grad.

After its compression and liquefaction at the location of Fractionation Facilities Ivanić Grad, CO_2 is transported by pipeline at high pressure (200 bar) to the injection wells of the Ivanić and Žutica fields, in quantities of 400,000 and 200,000 m^3/day, respectively. During the period of 25 years, which is the expected duration of the project, about 5×10^9 m^3 of CO_2 will be injected in the reservoirs of these fields. That will result in additional hydrocarbon production (3.4×10^6 t of oil and 599×10^6 m^3 of gas). Due to geological and physical conditions, about 50% of injected CO_2 will be permanently trapped in the reservoirs, while another 50% of CO_2 will be produced together with associated gas. Currently, the solution regarding the further use of CO_2, which will be extracted from associated gas at the location of the Compressor Station Žutica, is being developed. To implement the EOR project, it was necessary to carry out workover operations and construction modifications of existing wells. Keeping in mind corrosive features of CO_2, special attention was paid to the selection of surface and underground equipment.

According to Heidug et al. [26], CO_2-EOR practice can be modified to deliver significant capacity for long-term CO_2 storage. EOR expansion to storage of CO_2 can be achieved through at least four major activities: (1) additional site characterisation and risk assessment to evaluate the storage capability of a site, (2) additional monitoring of vented and fugitive emissions, (3) additional subsurface monitoring and (4) changes to field abandonment practices.

5. Geological storage complex and surrounding area characterisation

Potential sites for geologic storage are depleted oil and gas fields, deep saline formations and deep unmineable coal seams. According to EU Directive 2009/31/EC [7], the characterisation and assessment of the potential storage complex, including the cap rock and surrounding area, including the hydraulically connected areas, should be carried out in three steps according to best practices at the time of the assessment: (1) data collection, (2) building the three-dimensional static geological earth model and (3) characterisation of the storage dynamic behaviour, sensitivity characterisation and risk assessment (**Figure 5**).

Collecting data about the storage complex and the surrounding area is very important because it serves as a base for making their volumetric and three-dimensional (3-D) static earth model.

In the first step, for describing the storage complex, it is necessary to collect information about its characteristics. In the second step, based on the collected data and using computerised reservoir simulators, a three-dimensional static geological earth model of the candidate storage complex, including the cap rock and the hydraulically connected areas and fluids, is built. It characterises the storage complex.

In the third step, the characterisations and assessment of storage complex are based on dynamic modelling, comprising a variety of time-step simulations of CO_2 injection into the storage site using the three-dimensional static geological earth model(s) constructed during the second step. The simulations are based on altering parameters in the static geological earth model(s) and changing rate functions and assumptions in the dynamic modelling exercise. Any significant sensitivity should be taken into account during risk assessment.

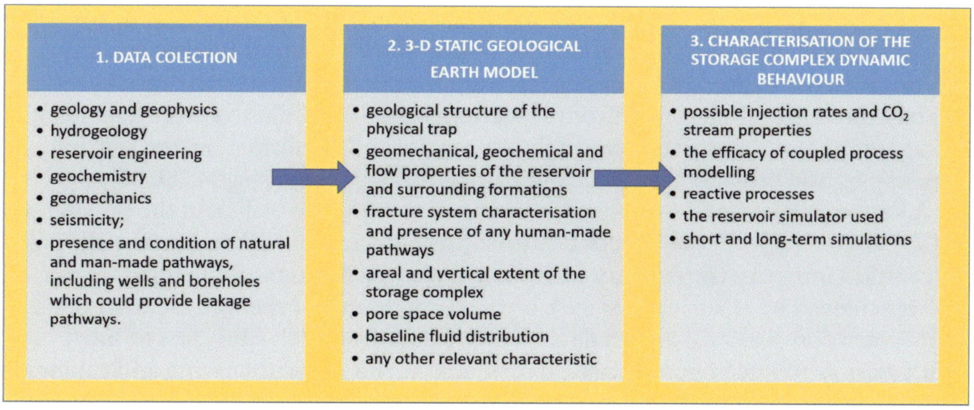

Figure 5.
The characterisation and assessment of the storage complex.

6. Potential CO_2 leakage pathways

The injected CO_2 could leak or migrate from CO_2 storage formation upwards (into upper rocks, aquifer or to atmosphere) if the following conditions are present: (a) CO_2 gas pressure exceeds capillary pressure and passes through siltstone, (b) free CO_2 leaks from siltstone into upper aquifer up the fault, (c) CO_2 escapes through a "gap" in the cap rock into a higher aquifer, (d) injected CO_2 migrates up the dip, increases reservoir pressure and permeability of fault, (e) CO_2 escapes via poorly plugged new or old abandoned wells, (f) natural flow dissolves CO_2 at CO_2/water interface and transports it out of closure and (g) dissolved CO_2 escapes to the atmosphere or into the ocean. **Figure 6** shows the migration paths of injected CO_2 from storage formation towards surface through a fracture in the cap rock, along fault zones and via poorly cemented active or abandoned wells.

The integrity of the cap rock is assured by an adequate fracture gradient and by sufficient cement around the casing across the cap rock and without a micro-annulus. The permeability and integrity of the cement will determine how effective it is in preventing leakage.

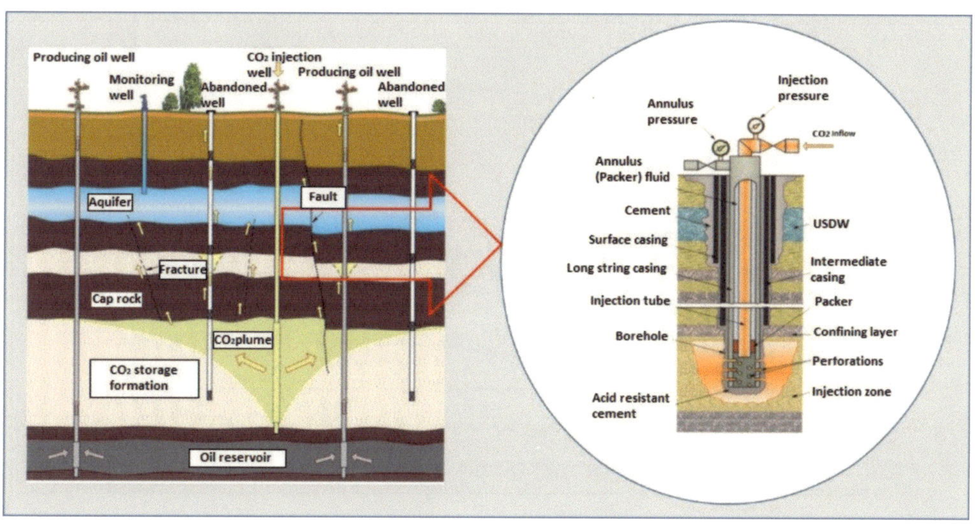

Figure 6.
Potential leakage pathways of injected CO_2 and CO_2 injection well design (modified after references [27, 28]).

Potential leakage pathways along an active injection well and/or an abandoned well include leakage: through deterioration (corrosion) of the tubing, around packer, through deterioration (corrosion) of the casing, between the outside of the casing and the set cement, through the deterioration of the set cement in the annulus (cement fractures), leakage in the annular region between the set cement and the formation, through the cement plug and between the set cement and the inside of the casing [4, 29, 30].

A key concept related to the performance of an injection well, and the prevention of CO_2 migration from the injection zone through an active or abandoned well, is its mechanical integrity (internal and external). Internal mechanical integrity of the well is achieved by ensuring that each of the components of the well is constructed using corrosion-resistant materials such as 316 stainless steel, fibreglass or lined (with glass reinforced epoxy, plastic or cement) carbon steel for casing and tubing. External mechanical integrity of the well is achieved by successful primary cementing operation with the use of CO_2-resistant cement, resulting in a cement sheath to bond and support casing and provide zonal isolation. The permeability and integrity of the set cement will determine its effectiveness in preventing CO_2 leakage.

6.1 CO_2 trapping mechanisms

The possibility of potential leaks of CO_2 is one of the largest barriers to large-scale CCS although well-selected storage sites are likely to retain over 99% of the injected CO_2 over 1000 years. Four different storage mechanisms keep the supercritical CO_2 securely stored inside the CO_2 storage formation: structural/stratigraphic (or physical) trapping, (2) solubility trapping, (3) residual trapping and (4) mineral trapping [1, 31]. The most important CO_2 storage mechanism during an injection process of several decades is structural/stratigraphic trapping. The other three mechanisms enable the trapping of CO_2 over a long period of time [1]. The effectiveness of geological storage depends on a combination of physical and geochemical trapping mechanisms. **Figure 7** presents four injection scenarios.

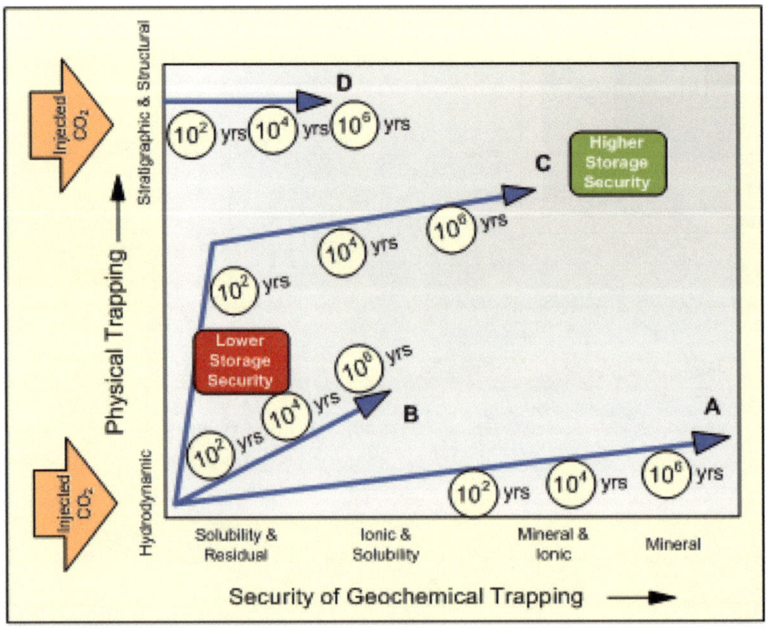

Figure 7.
The influence of a combination of physical and geochemical trapping mechanisms on CO_2 security storage (modified after reference [1]).

Injection scenarios A, B and C show injection into hydrodynamic traps, essentially systems open to lateral flow of fluids and gas within the injection formation. Scenario D represents injection into a physically restricted flow regime, similar to those of many producing and depleted oil and gas reservoirs. The level of security is proportional to the distance from the origin. Dashed lines are examples of million-year pathways.

As time passes and more CO_2 is injected, the more secure trapping mechanisms keep CO_2 in place, leading to increased security of storage [31].

7. Storage capacity

According to Bradshaw et al. [32], capacity calculation can be threefold, depending on the required category level: *theoretical, realistic and viable capacity*. *Theoretical capacity* considers whole reservoir pore space available for storage, or saline aquifer, which is saturated with salt water having maximum dissolved CO_2. In practice, different technical and economic restrictions prevent storage quantities to reach the level of theoretical capacity. *Realistic capacity* takes into consideration reservoir quality parameters (porosity, permeability, seal, depth, pressure, stress regimes, etc.) as important indications of technical viability. *Viable capacity* includes legal and regulatory limitations and considers social and environmental aspects of the selected location while connecting the CO_2 source with the nearest storage site.

Storage capacity can be generally expressed as the quantity of CO_2 that may be injected and stored in the geological layers.

According to the study of the Task Force for Review and Identification of Standards for CO_2 Storage Capacity Estimation of Carbon Sequestration Leadership Forum (CSLF), the regional CO_2 storage capacity in structural and stratigraphic traps (Eq. (1)) can be calculated using a residual water saturation [33, 34]:

$$V_{CO2t} = V_{trap} \cdot \Phi(1 - S_{wirr}) = A \cdot h \cdot \Phi\,(1 - S_{wirr}) \qquad (1)$$

where V_{CO2t}, theoretical storage volume CO_2 (m^3); V_{trap}, trap volume (m^3); Φ, average trap porosity (−); S_{wirr}, irreducible water saturation (−); A, trap area (m^2); h, average trap thickness (m).

Similar approach is used by the United States Department of Energy (DOE). It takes into account the porous space of the entire layer of saturated water and does not distinguish between CO_2 storage mechanisms. It takes into account the storage efficiency coefficient, which reflects the size of the space that can be filled with CO_2. The coefficient encompasses a wide variety of variables, ranging from petrophysical reservoir properties (porosity and permeability) to the sweep efficiency and effective porosity. According to the US DOE, for the regional salt water aquifers, the coefficient of storage efficiency is suggested to be 2% [35, 36].

The storage capacity of depleted hydrocarbon fields [Eqs. (2) and (3)] can be calculated from cumulative production and reserve data following the methodology described in [37].

$$M = \rho_{\cdot CO_{2r}} \left(R \cdot \mathrm{f} \cdot \mathrm{N} \cdot B_{\mathrm{fo}} - W_{\mathrm{i}} + W_{\mathrm{p}}\right) \qquad (2)$$

$$M = \rho_{\cdot CO_{2r}} \cdot R\mathrm{f}\,(1 - F_{\mathrm{ig}}) \cdot G \cdot B_{\mathrm{g}} \qquad (3)$$

where M, reservoir capacity for CO_2 storage (kg); ρ_{CO2r}, CO_2 density at reservoir conditions (kg/m^3); R_f, recovery factor (−); N, original oil in place (m^3); B_o, oil formation volume factor (−); W_i, water injection (m^3); W_p, water production (m^3); F_{ig}, gas injection (m^3); G, original gas in place (m^3) and B_g, gas formation volume factor (−).

Theoretical storage capacity obtained by these equations takes into account the estimated recoverable hydrocarbon reserves as the product of original hydrocarbon in place and recovery factor. For the effective capacity, it is necessary to consider some additional factors such as the macroscopic displacement efficiency, buoyancy, reservoir heterogeneity, water saturation, reservoir drive, etc.

Although the sweep efficiency has often been ignored in the case of depleted hydrocarbons fields, instead of the total amount, only 75% replacement of original oil or gas in place can be expected [38, 39].

The very first global assessment of CO_2 storage capacity was made back to the 1990s. Koide et al. [40, 41] assessed CO_2 storage capacity for deep saline aquifers on the level of 320×10^9 t. According to Van der Meer [42], it was estimated to 425×10^9 tons, calculation made by Ormerod et al. [43] was on the level of 790×10^9 t CO_2. Hendricks and Blok [44] reported storage capacity of 150×10^9 t, which was mainly related to depleted hydrocarbon reservoirs [25].

Preliminary estimation of CO_2 storage capacity for European deep aquifers and hydrocarbon reservoirs was done within the framework of the projects GESTCO, CASTOR and GeoCapacity, financed under the 5th and 6th Framework Program for Research and Technological Development [45]. In the case of deep aquifers, a simplified methodology based on a volumetric approach was applied, calculating with average values for layer thickness, temperature, pressure and porosity for each storage location. Storage assessment of hydrocarbon reservoirs used material balancing method, assuming that extraction of hydrocarbon releases certain pore volume available for CO_2 injection. The EU GeoCapacity project estimated CO_2 storage capacity to be on the level of 127 Gt, covering saline formations (97 Gt), hydrocarbon fields (20 Gt) and coal seams (1 Gt). The storage capacity was evaluated in 17 countries as sufficient at national level, while in one country (Norway), it was concluded that cross-border storage is possible. However, storage capacity was defined as "insufficient" in five countries [14].

7.1 CO_2 storage resources classification

The Society of Petroleum Engineers (SPE) published the document entitled *CO_2 Storage Resources Management System (SRMS)*, prepared by its subcommittee of the Carbon Dioxide Capture, Utilization and Storage Technical Section (CCUS), which establishes technically based capacity and resources evaluation standards [45]. This document is based on the SPE PRMS (*The Petroleum Resources Management System*), which is developed by SPE Oil and Gas Reserves Committee and used internationally within the petroleum industry for consistent and reliable definition, classification and estimation of hydrocarbon resources.

SPE CO_2 SRMS provides a consistent approach to estimating storable quantities of CO_2, evaluating development projects and presenting results within a comprehensive classification framework. The SRMS classification scheme is based on the accessible pore volume in a geologic formation in which CO_2 could be stored. It is intended for use in geologic formation completely saturated with brine such as saline formations or saline aquifers and depleted hydrocarbon fields without hydrocarbon production.

CO_2 storage resources are defined as the quantity (mass or volume) of CO_2 that can be stored in a geological formation and include all quantities of naturally occurring pore volume potentially suitable for storage within underground formations—*discovered* and *undiscovered* (accessible and inaccessible storage resources), as well as those quantities already used for storage (stored resources). The SPE storage resources classification system is shown in **Figure 8**.

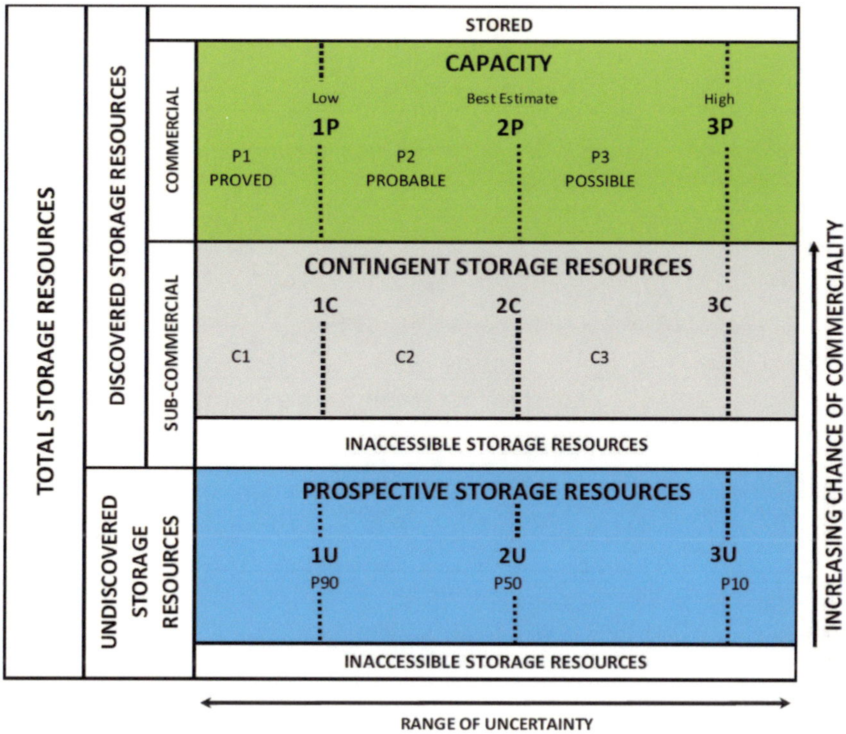

Figure 8.
CO_2 resources classification framework [45].

8. Risk assessment

The risks associated with underground CO_2 storage depend on many factors, including used infrastructure, type of reservoir dedicated to storage, geological characteristics of selected layers, cap rock and stratigraphic heterogeneity, geomechanical properties of rocks, existence of other wells, method of well abandonment experience, etc. EU CCS Directive is developed on the basis of a risk-based approach for safe storage and leakage. Therefore, it is necessary, before the application of CCS, to determine whether identified risks are acceptable. The significant risk of CO_2 leakage could not be permitted under the EU CCS Directive.

The risk assessment should comprise, among other things, hazard characterisation, exposure and effects assessment and risk characterisation. Characterisation of the hazard is carried out by characterising the potential leakage from the storage complex, as established by the dynamic modelling. It should cover the full range of potential operating conditions to test the security of the storage complex.

Many papers are published with the aim of assessing the risk of CO_2 storage, and various methodologies are currently applied to risk assessment of geological CO_2 storage (e.g., [3, 4, 28, 46–48]).

Figure 9 shows risk concept profiles for a large CCS project over time. The blue line represents a project with the pressure in storage formation increasing during CO_2 injection and decreasing after injection stops. The red line represents potential risk profile over time. The potential risk of failure and CO_2 leakage increases during the injection, and after the injection stops, it decreases. Secondary risk increases depend on local geochemical risks of transport processes.

Jewell and Senior [51] described scenarios and parameters for potential leakage from active (CO_2 injection well, observation well or water extraction well) and

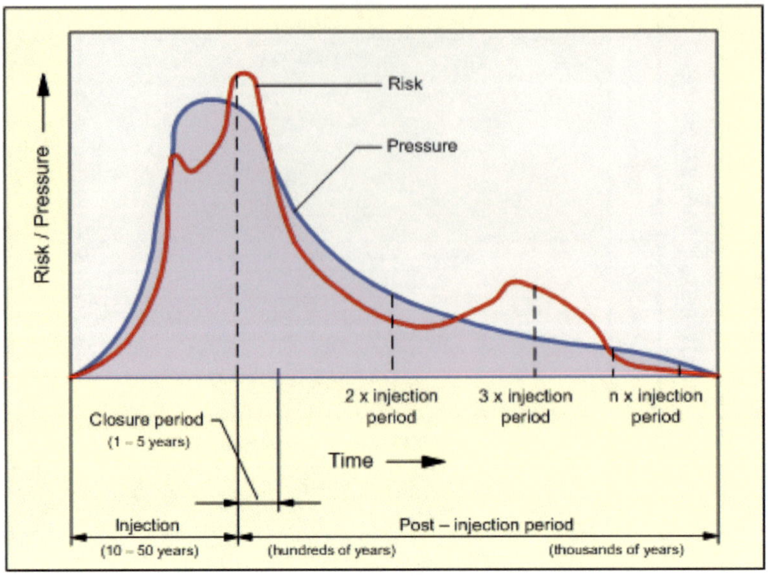

Figure 9.
Risk concept profiles for a large CCS project over time (modified after references [3, 49]).

Parameters	Scenarios			
	Active CO_2 injection well		Abandoned well	
	Low-level leakage: via CO_2 injection well	Worse case: blow out on CO_2 injection well after failure of initial well control activities	Low-level leakage: via abandoned well	Worse case: complete breakdown of abandonment plugs in old well
Probability of leakage	0.0001–0.001	0.00001–0.0001	0.0012–0.005	—
Potential CO_2 leakage rates (t/day)	0.1–10.0	5000.00	0.60–6.00	1000.00
Duration	0.5–20 years (until well abandoned)	3–6 months	1–100+ years	3–6 months
Potential amount of CO_2 leakage (t)	18–73,000.00	$(0.45–0.9) \times 10^6$	220–220,000.00+	90–180,000.00
% CO_2 stored (200 million tonnes case)	0–0.036	0.225–0.45	0.0001–0.1+	0.045–0.09
Remarks	Data represent the best efforts to represent leakage scenarios and risks in the North Sea for a storage scheme:			
	With 5 CO_2 injection wells, 20-year injection period and 200 million tonnes of stored CO_2.		With 6 abandoned wells, probability of leakage over 100 years and 200 million tonnes of stored CO_2.	

Table 1.
Scenarios and parameters for potential leakage from active and abandoned wells (modified after reference [50]).

abandoned wells as well as via primary cap rock and fault to assist in the development of a common understanding of CO_2 leakage and associated liabilities in the North Sea (**Tables 1** and **2**).

Parameters	Scenarios			
	Primary cap rock	Fault		
	Migration through primary rock	Low flux: vertical migration through existing faults	Moderate flux: vertical migration through existing faults	High flux: migration through fault activated and enhanced by injection
Probability of leakage	Negligible	Not calibrated—highly site specific		
Potential CO_2 leakage rates (t/day)	Very low flux rates	1–50	50–250	1500.00
Duration	100–1000 years to breakthrough	1–100 years for low flux; excludes remediation	1–5 years; includes remediation	1–5 years; includes remediation
Potential amount of CO_2 leakage (t)	Very low	$(0–1.8) \times 10^6$ (100-year flux); no remediation	$(0.018–0.46) \times 10^6$ including remediation	$(0.55–2.7) \times 10^6$ including remediation
% CO_2 stored (200 million tonnes case)	N/A	0–0.9	0.0009–0.23	0.275–1.37
Remark	—	Data represent the best efforts to represent leakage scenarios and risks in the North Sea if faults are present.		

Table 2.
Scenarios and parameters for potential leakage via primary cap rock and fault (modified after reference [50]).

In case of leakages or significant irregularities, the operator is obliged to immediately notify the competent authority and take the necessary corrective measures, including measures related to the protection of human health. The purpose of corrective measures is to prevent or stop the escape of CO_2 from the storage formation, to ensure safe geological storage and to manage the risks during the lifespan of the project and afterwards. According to the EU CCS Directive and EC Guidance Document 2, corrective measures include but are not limited to (1) limiting CO_2 injection rates or stopping injection and pressure buildup, (2) reducing the reservoir pressure by extracting CO_2 or water from the storage complex, close to an identified leakage area or applying peripheral extraction, (3) sealing areas of leakage such as identified fault or cap rock leakage pathways by injecting low permeability materials, creating a hydraulic barrier that stops CO_2 migration in sensitive areas by increasing the pressure in the above formations, (4) well remediation for active wells (for example, repair of wellhead, damaged tubing or collapsed casing; packer replacement, squeeze cementing and so on) and (5) well control, including killing the well by injecting heavy fluids and after that cementing the well or drilling a new well to intersect and plug the leaking well.

9. CO_2 injection monitoring

Monitoring of injection facilities, storage complex (including where possible the CO_2 plume) and, where appropriate, the surrounding environment present a very important part of the overall risk management strategy for geological storage projects. It should be based on a monitoring plan established according to the risk assessment analysis.

	Preoperational	Operational	Closure
Basic monitoring	**Monitoring program**		
	Well logs	—	—
	Wellhead pressure	Wellhead pressure	—
	Formation pressure	—	—
	Injection- and production rate testing	Injection- and production rates	—
	Atmospheric-CO_2 monitoring	Wellhead atmospheric-CO_2 monitoring	—
	Seismic survey	Seismic survey	Seismic survey
	—	Microseismicity	—
Enhanced monitoring	**Additions to the basic monitoring program**		
	—	Well logs	—
	—	—	Wellhead pressure
	CO_2-flux monitoring	Continuous CO_2-flux monitoring	CO_2-flux monitoring
	Gravity survey		
	Electromagnetic survey		
	Pressure and water quality above the storage formation		

Table 3.
Monitoring program for geologic storage of CO_2 *(modified after [51]).*

Benson et al. [52] provided examples of basic and enhanced programs that could be deployed for geologic storage of CO_2. They include preoperational, operational and closure monitoring program and could be used over the lifetime of a geologic storage project. Their application in practice will enable the implementation of the CO_2 injection project and increase security and reduce the risk of migration of injected gas, thus protecting the environment (**Table 3**).

The choice of monitoring technology should be based on best practice available at the time of the design.

The parameters to be monitored are identified so as to fulfil the purposes of monitoring. However, the monitoring plan should in any case include continuous or intermittent monitoring of (1) fugitive emissions of CO_2 at the injection facility; (2) CO_2 volumetric flow at injection wellheads; (3) CO_2 pressure and temperature at injection wellheads (to determine mass flow); (4) chemical analysis of the injected material and (5) reservoir temperature and pressure (to determine the CO_2 phase behaviour and state).

The monitoring plan should be updated if new CO_2 sources, pathways and flux rates or observed significant deviations from previous assessments are identified.

Post-closure monitoring is based on the information collected and modelled during the implementation of the monitoring plan.

10. Conclusions

Increment of greenhouse gases in the atmosphere is a direct consequence of industrial development. It manifests itself in rise of the average earth temperature being responsible for a series of unfavourable climate changes. CCS can help in mitigating climate changes through a distinctive huge sequestration capacity, which ensures global utilisation. Technology applicability and safety have been testing by several large- and small-scale demonstration projects currently under way.

Switching CCS technology from demonstration to commercial deployment depends on CO_2 market price. Although current value is not encouraging, more stringent emission reduction strategy (80–95% by 2050) will lead to commercial applications. However, besides emission reduction initiatives, there are many projects connected to EOR activities. Viability of such projects is strongly dependent on the oil price.

Since geological storage permanence is enabled by natural and engineered barriers functionality, there is a certain risk of migration of CO_2 from the storage formation. The potential leakage risk increases during injection phase, and with time, it decreases due to activation of different trapping mechanisms. Therefore, structural/stratigraphic trapping represents the most important CO_2 storage mechanism in the first storage period. The other mechanisms take over with storage life progressively. Mineral tapping of CO_2 is the safest mechanism, as CO_2 reacts with the reservoir rock minerals and remains permanently trapped.

Well-selected, designed and managed geological storage sites pose the risks comparable to those associated with current hydrocarbon recovery activities. Such risks, determined by leakage rates of less than 1% over thousands of years, are well below levels that could endanger public safety or environment. Nevertheless, for all CCS projects, a comprehensive monitoring, including baseline, operational and post-closure state, is mandatory.

Author details

Nediljka Gaurina-Međimurec* and Karolina Novak Mavar
Faculty of Mining, Geology and Petroleum Engineering, University of Zagreb, Zagreb, Croatia

*Address all correspondence to: nediljka.gaurina-medjimurec@rgn.hr

References

[1] Intergovernmental Panel on Climate Change (IPCC). Special Report on Carbon Dioxide Capture and Storage. Cambridge, UK: Cambridge University Press; 2005. p. 431

[2] Gaurina-Međimurec N, Novak Mavar K, Majić M. Carbon capture and storage (CCS): Technology, projects and monitoring review. The Mining-Geological-Petroleum Engineering Bulletin. 2018;**33**:1-15. DOI: 10.17794/rgn.2018.2.1

[3] Forbes SM, Verma P, Curry TE, Friedmann SJ, Wade SM. Guidelines for Carbon Dioxide Capture, Transport, and Storage. Washington, DC: World Resources Institute; 2008. p. 148

[4] Gaurina-Međimurec N, Pašić B. Design and mechanical integrity of CO_2 injection wells. The Mining-Geology-Petroleum Engineering Bulletin. 2011;**23**:1-8

[5] European Commission (EC). Communication on a Policy Framework for Climate and Energy from 2020 to 2030, COM(2014)0015; Brussels, Belgium. 2014. p. 20

[6] European Commission (EC). A Clean Planet for all: A European Strategic Long-Term Vision for a Prosperous, Modern, Competitive and Climate Neutral Economy, COM(2018)773; Brussels, Belgium; 2018. p. 25

[7] Directive 2009/31/EC of the European Parliament and of the Council of 23 April 2009 on the Geological Storage of Carbon Dioxide and Amending Council Directive 85/337/EEC, European Parliament and Council Directives 2000/60/EC, 2001/80/EC, 2004/35/EC, 2006/12/EC, 2008/1/EC and Regulation (EC) No 1013/2006. OJ L 140

[8] Directive 2003/87/EC of the European Parliament and of the Council of 13 October 2003 Establishing a Scheme for Greenhouse Gas Emission Allowance Trading within the Community and Amending Council Directive 96/61/EC. OJ L 275

[9] European Commission (EC). Report on the Functioning of the European Carbon Market, COM(2017)48 Final; Brussels, Belgium; 2017. p. 34

[10] Intergovernmental Panel on Climate Change (IPCC). Climate Change 2014: Synthesis Report. Contribution of Working Groups I, II and III to the Fifth Assessment Report of the Intergovernmental Panel on Climate Change. Geneva, Switzerland: IPCC; 2014. p. 151

[11] International Energy Agency (IEA). Energy Technology Perspectives 2016: Towards Sustainable Urban Energy Systems; Paris; 2016. p. 16

[12] Global CCS Institute. Annual Reportp; Cannberra, Australia; 2012. p. 40

[13] Kapetaki Z, Scowcrofta J. Overview of Carbon Capture and Storage (CCS) demonstration project business models: Risks and enablers on the two sides of the Atlantic. Energy Procedia. 2017;**114**:6623-6630

[14] Shogenovaa A, Piessensb K, Holloway S, Benthamc M, Martínezd R, Flornese KM, et al. Implementation of the EU CCS directive in Europe: Results and development in 2013. Energy Procedia. 2014;**63**:6662-6670

[15] Carbon Capture and Sequestration Technologies at Massachusetts Institute of Technology (MIT) [Internet]. Available from: https://sequestration.mit.edu [Accessed: 02 December 2018]

[16] Global CCS Institute. The Global Status of CCS: 2017; Australia; 2017. p. 82

[17] International Energy Agency (IEA). Technology Roadmap Carbon Capture and Storage; Paris; 2013. p. 59

[18] USA National Energy Technology Laboratory (NETL) [Internet]. Available from: https://www.netl.doe.gov [Accessed: 02 December 2018]

[19] Kheshgi HS, Thomann H, Bhore NA, Hirsch RB, Parker ME, Teletzke GF. Perspective on CCS costs and economics. SPE Economics & Management. 2012;**4**:24-30

[20] Global CCS Institute [Internet]. Available from: https://www.globalccsinstitute.com [Accessed: 02 December 2018]

[21] Zero Emissions Platform Database [Internet]. Available from: http://www.zeroemissionsplatform.eu/ [Accessed: 02 December 2018]

[22] CO_2 Stored Database [Internet]. Available from: http://www.co2stored.co.uk/home/index [Accessed: 02 December 2018]

[23] Office of Fossil Energy [Internet]. Available from: http://energy.gov/fe/science-innovation/oil-gas/enhanced-oil-recovery [Accessed: 02 December 2018]

[24] Shell [Internet]. Available from: http://www.shell.com/global/future-energy/going-underground/eor.html [Accessed: 02 December 2018]

[25] Novak K. Modeliranje površinskoga transporta i geološki aspekti skladištenja ugljikova dioksida u neogenska pješčenjačka ležišta Sjeverne Hrvatske na primjeru polja Ivanić (Surface Transportation Modelling and Geological Aspects of Carbon-Dioxide Storage into Northern Croatian Neogene Sandstone Reservoirs, Case Study Ivanić Field) [thesis]. University of Zagreb, Faculty of Mining, Geology and Petroleum Engineering; 2015

[26] Heidug W, Lipponen J, McCoy S, Benoit P. Storing CO_2 Through Enhanced Oil Recovery, Combining EOR with CO_2 Storage (EOR+) for Profit. International Energy Agency (IEA); 2015. p. 48

[27] Gaurina-Međimurec N, Pašić B. CO_2 underground storage and wellbore integrity. In: Matanović D, Gaurina-Međimurec N, Simon K, editors. Risk Analysis for Prevention of Hazardous Situations in Petroleum and Natural Gas Engineering. Hershey: IGI Global; 2013. pp. 169-217

[28] Gaurina-Međimurec N, Novak Mavar K. Depleted hydrocarbons reservoirs and CO_2 injection Wells-CO_2 leakage assessment. The Mining-Geology-Petroleum Engineering Bulletin. 2017;**2**:15-27. DOI: 10.17794/rgn.2017.2.3

[29] Gasda S, Bachu S, Celia MA. The potential for CO_2 leakage from storage sites in geological media: Analysis of well distribution in mature sedimentary basin. Environmental Geology. 2004;**46**(6-7):707-720

[30] Gaurina-Međimurec N. The influence of CO_2 on well cement. The Mining-Geology-Petroleum Engineering Bulletin. 2010;**22**:19-25

[31] International Energy Agency (IEA). Interaction of CO_2 Storage with Subsurface Resources. Report: 2013-08. UK: IEA Environmental Projects Ltd. (IEAGHG); 2013. p. 115

[32] Bradshaw J, Bachu S, Bonijoly D, Burruss R, Holloway S, Christensen NP, et al. CO_2 storage capacity estimation: Issues and development of standards. International Journal of Greenhouse Gas Control. 2007;**1**:62-68

[33] Carbon Sequestration Leadership Forum (CSLF). Phase I Final Report from the Task Force for Review and Identification of Standards for CO_2

Storage Capacity Measurement, CSLF-T-2005-9 15. Berlin, Germany; 2005. p. 16

[34] Carbon Sequestration Leadership Forum (CSLF). Phase II Final Report from the Task Force for Review and Identification of Standards for CO_2 Storage Capacity Estimation, CSLF-T-2007-04; Melbourne, Australia; 2007. p. 42

[35] U.S. Department of Energy (DOE). Carbon Sequestration Atlas of United States and Canada. Washington: Office of Fossil Energy; 2007. p. 86

[36] U.S. Department of Energy (DOE). Methodology for Development of Geologic Storage Estimates for Carbon Dioxide. Washington: Office of Fossil Energy; 2008. p. 36

[37] Bachu S, Bonijoly D, Bradshaw J, Burrus R, Holloway S, Christensen NP, et al. CO_2 storage capacity estimation methodology and gaps. Internationa Journal of Greenhouse Gas Control. 2007;**1**(4):430-443

[38] Holloway S, Vincent C, Kirk K. Industrial Carbon Dioxide Emissions and Carbon Dioxide Storage Potential in the UK, DTI Cleaner Fossil Fuels Programme Report. DTI/Pub URN 06/2027; UK; 2006. p. 47

[39] International Energy Agency (IEA). CO_2 Storage in Depleted Gas Fields. Technical Study, Report No. 2009/01; UK: IEA Greenhouse Gas R&D Programme. 2009. p. 121

[40] Koide H, Tazaki Y, Noguchi Y, Nakayama S, Iijima M, Ito K, et al. Subterranean containment and long-term storage of carbon dioxide in unused aquifers and in depleted natural gas reservoirs. Energy Conversion and Management. 1992;**33**(5/8):619-626

[41] Koide H, Tazaki Y, Noguchi Y, Iijima M, Ito K, Shindo Y. Underground storage of carbon dioxide in depleted natural gas reservoirs and in useless aquifers. Engineering Geology. 1993;**34**(3/4):175-179

[42] Van der Meer L. Investigations regarding the storage of carbon dioxide in aquifers in the Netherlands. Energy Conversion and Management. 1992;**33**(5/8):611-618

[43] Ormerod WG, Webster IC, Ausdus H, Reimer PWF. An overview of large scale CO_2 disposal options. Energy Conversion and Management. 1993;**34**(9/1):833-840

[44] Hendriks CA, Blok K. Underground storage of carbon dioxide. Energy Conversion and Management. 1993;**34**:949-957

[45] Schuppers JD, Holloway S, May F, Gerling P, Bøe R, Magnus C, et al. Storage Capacity and Quality of Hydrocarbon Structures in the North Sea and the Aegean Region. TNO-Report NITG 02-020-B; Utrecht: Netherland's Institute of Applied Geoscience; 2003. p. 54

[46] Society of Petroleum Engineers (SPE). CO_2 Storage Resources Management System. 2016. p. 43

[47] Benson SM, Hepple R, Apps J, Tsang CF, Lippmann M. Lessons Learned from Natural and Industrial Analogues for Storage of Carbon Dioxide in Deep Geological Formations: Report No. LBNL-51170. Berkeley, California, USA: Lawrence Berkeley National Laboratory; 2002. p. 135

[48] White CM, Straziser BR, Granite EJ, Hoffman JS, Pennline HW. Separation and capture of CO_2 from large stationary sources and sequestration in geological formations—Coalbeds and deep saline aquifers. Journal of the Air & Waste Management Association. 2003;**53**:645-715

[49] Le Guen Y, Meyer V, Poupard O, Houdu E, Chammas R, Oxand SAA.

Risk-Based Approach for Well Integrity Management over Long Term in a CO_2 Geological Storage Project. Paper SPE 122510 Presented at the 2009 SPE Asia Pacific Oil and Gas Conference and Exhibition; 4-6 August 2009; Jakarta, Indonesia; 2009. p. 13

[50] Benson SM. Potential Liabilities and Mitigation Strategies for CCS. Presentation at WRI Workshop [Internet]. 2007. Available from: http://pdf.wri.org/sally_benson_ccs_june5.pdfhttp://pdf.wri.org/sally_benson_ccs_june5.pdf [Accessed: 02 December 2018]

[51] Jewell S, Senior B. CO_2 Storage Liabilities in the North Sea, an Assessment of Risks and Financial Consequences, Summary Report. UK: Department of Energy & Climate Change (DECC); 2012. p. 29

[52] Benson SM, Hoversten M, Gasperikova E, Haines M. Monitoring protocols and life-cycle costs for geological storage of carbon dioxide. In: Proceedings of the 7th International Conference on Greenhouse Gas Control Technologies; 5 September 2004. Vancouver, Canada: Elsevier; 2005. pp. 1259-1265